SpringerBriefs in Applied Sciences and Technology

Forensic and Medical Bioinformatics

Series Editor

Amit Kumar, Dwarka Venkat Sai Nagar Colony, Munaganoor, Hayatnagar, BioAxis DNA Research Centre Private Ltd., Hyderabad, India

The books of this series are submitted to SCOPUS, Google Scholar and Springerlink

Forensic and Medical Bioinformatics (FMB) series is a platform to bring interdisciplinary Technology driven Discovery and Disruptive content useful for Industry professionals, Researchers and Academicians and Advance Research students. We solicit contributions related to novel Research and innovation, Theory, applications, their technical elements and practical approaches related to

- Bio inspired technologies and Engineering in Biotechnology and Medical Sciences, Translational engineering
- Machine Learning and Artificial Intelligence in Health, Bioinformatics and Computational Biology
- Biosensors, Circuit and Systems Bio engineering, Communication technologies, Therapeutic systems and technologies, Biorobotics, Biomedical Signal Processing, BioMEMS, Neuro engineering, Biomedical circuits and systems
- Forensic Science and related laws, Case Studies and Scientific aids to investigation, Cyber Security, Bio Image Processing and Security.

To submit your proposal or show your interest for this series please contact: amit.kumar@dnares.in OR loyola.dsilva@springer.com

Robert Trybulski

Biomarker-Guided Physical Recovery Interventions

Microvascular and Neuromechanical Mechanisms in Sports Recovery

Robert Trybulski
Provita Medical Centre
Żory, Poland

Faculty of Medicine
Katowice Business University
Katowice, Poland

ISSN 2191-530X ISSN 2191-5318 (electronic)
SpringerBriefs in Applied Sciences and Technology
ISSN 2196-8845 ISSN 2196-8853 (electronic)
SpringerBriefs in Forensic and Medical Bioinformatics
ISBN 978-981-92-0630-8 ISBN 978-981-92-0631-5 (eBook)
https://doi.org/10.1007/978-981-92-0631-5

This Springer imprint is published by the registered company Springer Nature Singapore Pte Ltd.
The registered company address is: 152 Beach Road, #21-01/04 Gateway East, Singapore 189721, Singapore

Competing Interests The author has no competing interests to declare that are relevant to the content of this manuscript.

Praise for *Biomarker-Guided Physical Recovery Interventions*

"As a clinician, researcher, and professor in physiotherapy, with long-standing academic work centered on pain, manual therapy, dry needling, and therapeutic exercise, I read Robert Trybulski's book with genuine interest and appreciation. What immediately stands out is that this is not a superficial book about recovery techniques, but a serious scientific effort to explain physical recovery as a measurable biological process. In a field where practice has often advanced faster than conceptual clarity, this short book offers exactly the kind of mechanistic and evidence-oriented perspective that is needed.

What I find especially valuable is the breadth and organization of the content. The book begins by framing recovery in terms of skeletal muscle microdamage, inflammatory signaling, microvascular regulation, pain modulation, and neuromechanical adaptation and then develops these foundations through dedicated chapters on reactive hyperemia, endothelial and autonomic responses, edema control, metabolite clearance, and objective biomarkers such as muscle tone, stiffness, and elasticity. From there, it addresses the principal intervention families, cold and heat therapies, compression, massage, lymphatic drainage, dry needling, and other neuromodulatory approaches, not as isolated methods, but as physiological stimuli with specific mechanisms, indications, and dose-related effects. This gives the proposal unusual coherence and makes the text both scientifically substantial and highly relevant for clinical interpretation.

I also particularly appreciate the translational ambition of the book. Robert does not merely review interventions descriptively. He proposes a biomarker-guided framework for understanding and designing recovery protocols according to variables such as temperature, pressure, time, and mechanical load. That approach is extremely important. One of the major limitations in the literature on post-exercise recovery has been the lack of standardization in treatment parameters, outcome selection, and interpretation of physiological effects. By focusing on reproducibility, endpoint selection, and clinically meaningful biological signals, this book has the potential to provide a more rigorous structure for both researchers and practitioners working in sports physiotherapy, rehabilitation, and performance settings.

From a personal perspective, I consider Robert Trybulski particularly well qualified to write this book. His scholarly trajectory shows a sustained and focused engagement with exactly the issues this volume addresses: microcirculatory responses, compression-based recovery, cold and contrast therapies, dry needling, muscle biomechanical properties, and the integration of physiological assessment into applied recovery practice. His profile is especially convincing because it brings together experimental research, clinical reasoning, and experience with high-performance athletes. That combination is not common, and it gives the project both scientific credibility and practical authenticity.

For all of these reasons, I believe this book can become an important reference for readers interested in a more precise, physiology-based understanding of recovery interventions. It will be highly valuable for physiotherapists, sports medicine professionals, rehabilitation specialists, researchers, and performance practitioners who want to move beyond generic recommendations and toward scientifically grounded decision-making."

—César Fernández de las Peñas, PT, Ph.D., *Professor of Physiotherapy, Universidad Rey Juan Carlos, Madrid, Spain*

Contents

Chapter 1
Recovery as a Biological Process

Abstract This chapter frames recovery as an evidence-based, multiscale biological process that links short-term performance restoration to longer-term adaptation, rather than a single rested versus tired state. It explains why recovery time courses differ across exercise types and physiological levels (metabolic, neuromuscular, microvascular, inflammatory, and perceptual), and why a single marker rarely captures the dominant limiter of readiness. It then details how unaccustomed or high-strain loading can disrupt sarcomeres and excitation–contraction coupling, initiating repair and remodeling cascades that may transiently impair force, range of motion, and function. The chapter connects inflammatory signaling to tissue repair, emphasizing regulated leukocyte responses and active resolution as prerequisites for effective regeneration and training adaptation. Finally, it shows how symptoms (e.g., DOMS, perceived tightness) and mechanical constructs (e.g., tone, stiffness) emerge from interacting neural, connective-tissue, and fluid-dynamic mechanisms, informing what to measure, and how to interpret it, in real athletes.

1.1 Recovery Is a Multiscale Process

Recovery is the time-dependent return of an athlete's functional capacity after stress, constrained by the slowest recovering limiting subsystem for the next task [1]. Recovery is therefore inherently task-specific, because the limiter for repeated sprints can differ from the limiter for maximal strength, technical precision, or endurance output [2]. A practical consequence is that "fully recovered" cannot be defined by a single sensation or biomarker without specifying the desired performance and context [1].

Highlights

- *Recovery must be defined relative to the next performance demand, not as a generic state.*

R. Trybulski, *Biomarker-Guided Physical Recovery Interventions*,
SpringerBriefs in Forensic and Medical Bioinformatics,
https://doi.org/10.1007/978-981-92-0631-5_1

- *Different physiological systems recover on different time scales after the same session.*
- *Monitoring must target the likely bottleneck for the upcoming task.*

1.1.1 "Fast" Dynamics: Metabolic–Ionic Restoration (Minutes to Hours)

Fast recovery dynamics are the mechanisms that restore energy buffering, membrane excitability, and short-term contractile function quickly enough to matter within the same session or same day [1]. The most time-critical energetic buffer is the phosphocreatine (PCr) system, because PCr resynthesis during recovery is tightly linked to oxidative ATP production and predicts repeated-bout power restoration in high-intensity work [3]. In humans, PCr recovery kinetics show rapid components on the order of ~ tens of seconds and slower components on the order of minutes, implying that partial readiness can return quickly while full energetic normalization may take longer [4]. Because PCr resynthesis is influenced by the metabolic milieu and recovery conditions, strategies that alter oxygen delivery, perfusion, or acid–base balance can shift the rate of rapid energetic restoration [5].

A second fast limiter is peripheral contractile inhibition driven by intramuscular perturbations such as inorganic phosphate (Pi) and hydrogen ion (H^+), which can reduce force and Ca^{2+} sensitivity at the cross-bridge level [6]. Mechanistic models and experimental data support that elevated Pi can impair force by altering cross-bridge kinetics and can also contribute to reduced Ca^{2+} release from the sarcoplasmic reticulum during fatigue [7]. Therefore, an athlete can regain freshness quickly when the primary limiter was acute metabolite/ionic disturbance, yet still show reduced output when the limiter is excitation–contraction coupling disruption (which is not purely metabolic) [8].

Practical Box

If a session includes unaccustomed eccentrics, early weakness can reflect excitation–contraction coupling impairment rather than only metabolite fatigue, so subjective recovery can be misleading within hours. This helps explain why athletes sometimes report acceptable sensations post-session yet show depressed force or power the next day after eccentric-biased loading.

Highlights

- *PCr resynthesis is a central fast recovery lever for repeated high-intensity performance, because it is coupled to oxidative ATP provision.*
- *Pi/H^+ related mechanisms can reduce force quickly via cross-bridge and Ca^{2+} handling effects, making fatigue partly a biochemical-mechanical phenomenon.*
- *Fast recovery can look complete subjectively while mechanical output remains limited when excitation–contraction coupling is impaired.*

1.1.2 "Intermediate" Dynamics: Autonomic and Central Regulation (Hours to > 24 h)

Intermediate dynamics are the mechanisms that restore the internal environment for high-quality training on the next day, especially substrate stores, autonomic balance, and neural readiness [2]. Glycogen restoration is a central intermediate limiter for endurance and mixed-modal sports, because muscle glycogen availability constrains subsequent exercise capacity and training quality [9]. The glycogen resynthesis trajectory is biphasic, with an early window (approximately 0–4 h) in which carbohydrate provision strongly supports rapid resynthesis, and a later window (4–24 h) in which total carbohydrate intake is the dominant determinant [9]. Empirically, recovery nutrition patterns can produce markedly different glycogen resynthesis rates across the first hours after exercise, which is one reason why two-a-day training is nutrition-sensitive [10].

Practical Box

After a glycogen-depleting session, inadequate carbohydrate intake can leave muscle glycogen incompletely restored at 24 h, which can reduce the quality of the next day's training [11]. After a very intense interval session, HRV may remain perturbed into the next day, suggesting that systemic readiness can lag behind local feelings [12].

Autonomic recovery is another intermediate domain because parasympathetic reactivation and sympathetic withdrawal can remain perturbed after high-intensity exercise even when local muscle sensations normalize [12]. Studies of post-exercise heart rate variability (HRV) show intensity-dependent recovery kinetics, indicating that higher-intensity work tends to prolong autonomic disturbance compared with lower-intensity sessions [12]. From a mechanistic perspective, autonomic disturbance matters because it reflects central-peripheral integration of stress load, and it can track readiness changes that are not captured by local muscle mechanical measures alone [13].

Practical Box

In highly trained endurance athletes, autonomic disturbance is small after prolonged work below the first ventilatory threshold, but HRV recovery is clearly delayed after higher-intensity sessions. This creates a practical rule: if intensity crosses key thresholds, autonomic markers may remain altered even when legs feel subjectively recovered [14].

Hormonal stress responses also sit in the intermediate window, because cortisol dynamics can peak during or after strenuous exercise and interact with substrate mobilization and recovery physiology [15]. Endocrine markers like testosterone-to-cortisol ratio have been discussed as indicators of training strain rather than definitive diagnostics, reinforcing that endocrine signals should be interpreted as context-dependent stress readouts [16].

Sleep is a high-leverage intermediate regulator because inadequate sleep impairs performance and recovery and can modify inflammatory and hormonal responses to training stress [17]. For example, sleep loss can elevate inflammatory mediators such as interleukin-6 and degrade neuromuscular coordination and cognitive performance relevant to sport execution [18].

Practical Box

Athletes often face sleep disruption around travel and competition schedules, and evidence syntheses indicate that sleep disturbance is linked to impaired recovery and altered inflammatory/hormonal milieu [19]. In that context, intermediate dynamics can become the bottleneck even if the prior session was not mechanically damaging, because systemic regulation is compromised [17].

Highlights

- *Glycogen restoration is a principal limiter for next-day endurance and mixed-sport capacity, and its drivers differ between early (0–4 h) and later (4–24 h) recovery.*
- *Autonomic recovery is intensity-sensitive and can lag behind subjective recovery, so HRV/HRR are best treated as systemic readiness indicators.*
- *Sleep is a biological recovery intervention because it shapes neurocognitive output and inflammatory/hormonal regulation after training stress.*

1.1.3 "Slow" Dynamics: Tissue Remodeling, Microvascular–Fluid Balance, and Perception (Days)

Slow recovery dynamics are the mechanisms that restore or remodel tissue structure and function after mechanically stressful loading, and these processes typically unfold across days [20]. Eccentric-biased exercise can generate primary mechanical disruption and secondary damage processes (including Ca^{2+} dysregulation and ROS-related pathways), which extends recovery beyond metabolic timescales [21]. A central slow limiter is excitation–contraction coupling disruption driven by damage to membrane systems involved in Ca^{2+} release and transmission of the activation signal [22]. This mechanism explains why force can remain depressed for days after eccentric damage even when metabolite clearance is complete [23].

Immune and inflammatory signaling are essential slow-domain processes because sterile inflammation coordinates debris clearance and subsequent regeneration, and the timing of resolution influences functional restoration [20]. Exercise induces time-varying cytokine responses, and reviews emphasize that these responses are phase-structured rather than static, which makes timing central for interpretation [24]. Muscle regeneration and remodeling engage myogenic programs including satellite cell activation, with evidence that satellite cell activity increases after damaging eccentric work and evolves across 24–72 h windows [25].

Practical Box

DOMS commonly follows an inverted-U time course and often peaks 24–72 h after novel eccentric work, which aligns with delayed sensitization and inflammatory dynamics rather than acute metabolite accumulation [26]. This is why day-2 soreness can be the critical limiter for technical quality and range of motion, even when day-0 feels manageable [27].

Slow dynamics also include extracellular matrix (ECM) and collagen turnover, which matter because ECM structure influences force transmission and passive mechanical behavior often perceived as stiffness [28]. Human studies demonstrate that collagen protein synthesis in tendon and muscle can rise within hours and peak around 24 h after strenuous exercise, indicating that connective tissue remodeling is a measurable recovery component [28]. Because collagen and myofibrillar synthesis have distinct trajectories and magnitudes, perceived "tightness" or measured stiffness can reflect ECM-related remodeling rather than only muscle fiber damage [28].

Practical Box

Muscle protein synthesis can remain elevated at 24–48 h after resistance exercise, indicating ongoing remodeling even as symptoms improve [29]. Collagen synthesis can also remain elevated beyond early recovery, supporting the idea that connective tissue and ECM remodeling may continue after pain decreases [28].

Adaptation signaling can initiate rapidly but expresses as slow structural change, as illustrated by mitochondrial biogenesis regulation where PGC-1α transcription can peak within hours after endurance exercise while phenotypic changes accrue over repeated exposures [30]. Accordingly, intensified training blocks can alter molecular responsiveness to an acute bout, showing that recovery capacity and adaptation signaling are themselves plastic and training-history dependent [31].

Highlights

- *Eccentric damage prolongs recovery via excitation–contraction coupling and membrane/Ca^{2+} handling disruption, not just soreness.*
- *Inflammatory signaling is phase-structured and resolution-dependent, so timing determines whether a marker reflects productive repair or delayed recovery.*
- *Remodeling includes both myofibrillar and collagen/ECM synthesis, which can shift stiffness and function over days even if pain has declined.*

1.1.4 The Coupling Rule

The three timescales are coupled because fast energetic restoration sets immediate performance capacity, intermediate systemic regulation sets next-day readiness, and slow tissue remodeling sets the ceiling for durable adaptation and injury risk [1]. A practical implication is that the dominant limiter can shift over time after the same session, moving from metabolite/PCr constraints (minutes) to glycogen/autonomic constraints (hours) to tissue/immune/ECM constraints (days) [2]. This mechanistic shift explains why different recovery tools can appear effective depending on when they are applied and which limiter is active, which is the conceptual bridge to later chapters on modality selection and biomarker guidance [20].

Practical Box

If repeated-sprint output is the next constraint within minutes, PCr kinetics and short-term oxidative recovery capacity are likely central, so interventions should be judged against repeated-bout power restoration. If tomorrow's training quality is the constraint, glycogen restoration and autonomic recovery become more explanatory, so nutrition and systemic regulation markers deserve priority. If performance must be preserved across multiple days after eccentric loading, the slow domain (excitement-contraction coupling disruption, immune resolution, ECM remodeling) becomes the primary explanatory layer, so soreness alone is insufficient to judge readiness.

1.2 Load-Induced Microdamage

Exercise-induced muscle damage (EIMD) is a syndrome of transient strength loss, soreness, swelling, and reduced function that is most consistently triggered by unaccustomed or high-dose eccentric contractions [32]. In practice, the earliest and most sensitive functional signature of EIMD is the immediate post-exercise drop in force-generating capacity, which often precedes peak soreness by 24–72 h [20]. The magnitude and time course of these symptoms depend strongly on intensity, duration, muscle length/joint angle, and the muscle group stressed, which is why same session, different soreness is a common athlete experience [20].

Initiation: Non-uniform Sarcomere Strain and Focal Structural Failure

A dominant mechanical explanation is that active lengthening produces non-uniform sarcomere strain, where weaker sarcomeres are stretched to longer lengths and become mechanically unstable [33]. This concept is formalized in the popping sarcomere hypothesis, which predicts focal disruption at longer muscle lengths and a rightward shift in the muscle's length–tension characteristics after damaging eccentric work [33]. Consistent with this, eccentric damage is often associated with increased passive tension and altered mechanical behavior, reflecting injury-related changes to structural elements and the muscle's viscoelastic environment [34].

Secondary Damage: Excitation–Contraction Coupling, Ca^{2+} Dysregulation, and Proteolysis

A mechanistic reason strength can drop substantially with relatively small visible fiber disruption is excitation–contraction (E–C) coupling failure, meaning the action potential-to-Ca^{2+} release process is impaired [35]. Classic work shows that after eccentric contractions, force loss can be disproportionate to overt structural tearing, supporting E–C coupling disruption as an early limiting step for performance [35]. Perturbations to membrane systems involved in Ca^{2+} handling (including junctional structures that support triad function) are repeatedly implicated in the post-eccentric reduction in Ca^{2+} release and contractile performance [36]. Downstream, Ca^{2+} dysregulation can activate Ca^{2+}-dependent proteases (e.g., calpains) and amplify cytoskeletal and membrane-associated injury, forming a plausible secondary damage loop [37].

Why Biomarkers and Soreness Can Mislead Practitioners

Blood markers such as creatine kinase (CK) can rise after EIMD, but inter-individual variability is large and CK does not provide a one-to-one readout of functional impairment or tissue-level mechanisms in a given athlete [38]. This is one reason "high CK with good function" and "low CK with poor function" are both observed in real sport contexts, especially when sampling time is inconsistent with the individual's peak [38].

Practical Box

Downhill running, deceleration-heavy field sessions, or eccentric-emphasis strength training can cause immediate reductions in strength and power that are larger than the athlete expects from soreness alone. Soreness typically peaks later than strength loss, so "I don't feel sore yet" is not evidence that neuromuscular function is fully recovered [27]. Practically, this timing mismatch explains why sprint speed, change-of-direction quality, and jump outputs can be compromised before the athlete reports high pain.

Highlights

- *EIMD is best conceptualized as a coupled mechanical–physiological process, not merely "microtears."*
- *Early performance loss is strongly linked to E–C coupling impairment and Ca^{2+} handling deficits.*
- *Soreness and blood biomarkers can lag behind (or diverge from) functional status, so they are insufficient as stand-alone recovery indicators.*

1.2.1 Force Transmission Architecture

Skeletal muscle force is transmitted not only longitudinally through tendons but also laterally through the fiber cytoskeleton to the sarcolemma and extracellular matrix (ECM), distributing stress across the tissue [39]. The costamere is a key structural node that mechanically links sarcomeres to the sarcolemma and ECM and is therefore positioned to concentrate and sense mechanical load during contractions [40]. Because eccentric contractions impose high force while the muscle is lengthening, structures responsible for mechanical stabilization and lateral force transfer become plausible "weak links" for initiating microdamage [37].

Costameres, Dystrophin–Glycoprotein Complex, and Membrane Stability

The dystrophin–glycoprotein complex (DGC) is a major trans-sarcolemmal linkage between the actin cytoskeleton and laminin in the ECM, supporting membrane stability under mechanical stress [41]. Disruption of DGC structure is associated with contraction-induced membrane fragility and injury susceptibility, illustrating how "connective" proteins can be primary determinants of damage risk under load [42]. Experimental evidence that lateral force transmission is impaired when dystrophin-related structures are compromised reinforces the idea that lateral pathways materially contribute to whole-muscle mechanics [43].

Integrins as Load-Bearing Mechanosensors with Signaling Consequences

Integrins (α7β1 in skeletal muscle) contribute to muscle–ECM coupling and are proposed to act as mechanosensors that translate load into intracellular signaling relevant to remodeling [44]. This matters for an evidence-based "theory-to-practice" framework because the same structures that protect against mechanical disruption also participate in the adaptive signaling that follows training stress [44].

Titin and Intermediate Filaments: Viscoelastic Control and Strain Buffering

Titin contributes substantially to passive force, stabilizes sarcomeres, and is increasingly framed as a structure with both mechanical and signaling roles in muscle [45]. Desmin-based intermediate filaments help integrate myofibrils and distribute stress, meaning cytoskeletal integrity can influence how local strain becomes global dysfunction [46]. From a practical standpoint, increases in passive stiffness after damaging exercise can reflect altered behavior of these non-contractile elements and the surrounding ECM, not only changes in tone [39].

Practical Box

After eccentric work, athletes may report a "tight" or "restricted" feeling that is plausibly linked to changes in passive mechanical properties and tissue fluid dynamics, not only pain perception. Imaging work in delayed onset muscle soreness (DOMS) shows that edema-like changes and stiffness can be tracked over days, supporting the idea that mechanical and fluid components evolve over time [47].

Highlights

- *Muscle is a force-transmission network in which costameres, DGC, integrins, titin, and ECM determine where stress concentrates.*

- *Structural "connectors" are simultaneously protective (stability) and instructive (mechanotransduction), linking damage risk to adaptation potential.*
- *Post-exercise stiffness can reflect changes in passive structures and tissue environment, so stiffness = tone is an incomplete model.*

1.2.2 The Repeated-Bout Effect (RBE) as a Biological "Memory" of Damage

A single eccentric bout can confer protection against subsequent bouts, reducing soreness, swelling, and force loss, which is termed the repeated-bout effect [48]. Proposed mechanisms include neural adaptations, strengthened cytoskeletal/ECM structures, and cellular remodeling that reduces disruption for the same external load [48]. In practice, the RBE explains why early-season sessions feel disproportionately damaging and why novelty is a predictable risk factor for marked DOMS [27].

Practical Box

If an athlete introduces Nordic hamstring eccentrics after a long absence, soreness and weakness are expected to be higher on the first exposures than after repeated exposures [48]. The same absolute workload can therefore have different recovery costs depending on novelty and recent exposure history [48].

Highlights

- *The RBE is a robust protective adaptation after eccentric exposure.*
- *"Novel load" is biologically meaningful because it predicts larger damage responses.*
- *Programming should treat novelty as a controllable recovery stressor.*

1.3 Inflammatory Signaling and Repair

Inflammation after muscle injury is not a nonspecific "bad thing," but a coordinated biological program that can determine whether regeneration succeeds or fibrosis and prolonged dysfunction dominate [49]. Following damaging exercise, intramuscular immune events can occur even when systemic markers are subtle, which helps explain why blood-only monitoring may miss meaningful local biology [20]. Human biopsy studies after eccentric exercise demonstrate changes in immune-related markers and cell presence within muscle, supporting the relevance of local immune dynamics in real-world exercise contexts [50].

Early Phase: Neutrophils and the "Damage–Repair Trade-Off"

Neutrophils often dominate the earliest stages of acute inflammation and can contribute to both debris handling and additional tissue disruption via oxidants and proteases [51]. In exercise-induced muscle injury, the presence and magnitude of neutrophil infiltration has been debated, but systematic evaluation indicates it is frequently observed across models and conditions [52]. Animal data show that neutrophils can exacerbate injury and influence macrophage infiltration, illustrating how early inflammatory "dose" can shape downstream repair [53].

Mid-to-Late Phase: Macrophage Heterogeneity and Regeneration Control

Macrophages are functionally heterogeneous and appear in spatiotemporal patterns that regulate debris clearance, resolution of inflammation, and satellite cell behavior during repair [54]. A commonly described sequence is early pro-inflammatory macrophage activity followed by a shift toward pro-resolving phenotypes that support regeneration and remodeling [55]. Because macrophages directly influence the muscle stem-cell niche, the inflammatory response should be understood as part of the regeneration machinery rather than an "after-effect" of damage [54].

Practical Box

When an athlete repeats high-eccentric load sessions too close together, they may stack neuromuscular impairment on top of an unresolved inflammatory–repair phase, increasing performance variability. Conversely, some inflammatory signaling is integral to remodeling, so the practical aim is usually not eliminate inflammation, but dose stress so repair resolves on schedule.

Highlights

- *Inflammation is a regulated repair program that can enable regeneration or contribute to prolonged dysfunction depending on context and magnitude.*
- *Neutrophils and macrophages can both help and harm, so timing and load-management are central to practical recovery strategy.*
- *Muscle biopsies support that local immune dynamics matter even when systemic markers are unremarkable.*

1.3.1 Sterile Inflammation Cascade

After mechanical or metabolic stress, damaged muscle can release or expose danger signals (damage-associated molecular patterns; DAMPs) that help the immune system discriminate "healthy self" from "damaged self." [56]. These signals are sensed by pattern recognition receptors (PRRs), including toll-like receptors (TLRs), which can regulate macrophage phenotypes and thereby influence regeneration outcomes [57]. In this framework, inflammation is initiated without infection (sterile inflammation), driven by endogenous molecules and ECM fragments generated during tissue stress and disruption [56].

HMGB1 and TLR4 as Examples of "Danger–Repair" Signaling

High mobility group box 1 (HMGB1) is a representative DAMP that can modulate skeletal muscle regeneration by acting on myoblast and endothelial cell functions [58]. TLR signaling can be "friend and foe," because it can drive necessary inflammatory activation while also risking excessive or mis-timed inflammation that impairs repair [57]. Experimental injury models demonstrate that TLR4 signaling can influence muscle damage and repair dynamics, indicating that innate sensing pathways can shape tissue outcomes [59].

From Recruitment to Resolution: Why Phenotype Switching Matters

A central concept is that effective regeneration requires not only immune cell recruitment but also timely transition toward resolution programs that limit collateral damage and enable rebuilding [55]. Macrophage-mediated regulation of satellite cell proliferation and differentiation is repeatedly highlighted as a core mechanism linking immune signaling to restored muscle structure and function [54]. When inflammatory signaling becomes excessive or prolonged, the risk of maladaptive remodeling (including fibrosis) increases, which is why repair quality is as important as repair speed [60].

Practical Box

Interventions that blunt symptoms may alter immune signaling pathways that are also involved in regeneration, so symptom reduction is not automatically equivalent to improved tissue repair. A mechanistic approach evaluates whether an intervention plausibly supports timely resolution and effective remodeling rather than only reducing soreness [61].

Highlights

- *Sterile inflammation begins with DAMP sensing by PRRs (e.g., TLR pathways), linking microdamage to immune activation.*
- *HMGB1 and TLR4 illustrate how "danger signals" can participate in both injury responses and regenerative processes.*
- *The practical target is a well-timed inflammatory sequence that clears debris and transitions into resolution to support high-quality remodeling.*

1.3.2 Resolution Biology

Resolution of inflammation is an active, regulated process mediated by specialized pro-resolving lipid mediators and coordinated cellular programs [62]. Failure of timely resolution can prolong pain, swelling, and functional impairment even if the initial damage magnitude was moderate [20]. Thus, recovery strategies should aim to support appropriate resolution without blunting necessary remodeling signals [2].

Practical Box

Pharmacologic suppression of inflammatory pathways can alter muscle repair biology in some contexts, illustrating that symptom reduction and adaptation are not always aligned endpoints [63].

Highlights

- *Resolution is an active biological program, not passive decay of inflammation.*
- *Persistent inflammation can extend functional deficits beyond the initial insult.*
- *Recovery choices should be mechanism-targeted and goal-dependent (performance vs. remodeling).*

1.4 Microvascular Regulation and Fluid Balance

Blood flow to contracting muscle (exercise hyperemia) is regulated by integrated local and systemic mechanisms that match O_2 delivery to metabolic demand during work [64]. Nitric oxide and vasodilating prostaglandins contribute meaningfully to exercise hyperemia in humans, and combined inhibition reduces skeletal muscle blood flow during exercise [65]. During exercise, contracting muscle can blunt sympathetic vasoconstriction (functional sympatholysis), helping preserve perfusion in active tissue despite high sympathetic drive [66]. ATP is one candidate signal that can attenuate sympathetic vasoconstriction and contribute to vasodilation through pathways that include NO and prostaglandins [67].

Why Perfusion Mechanisms Matter for "Recovery," Not Only Performance

Reactive hyperemia (reperfusion after brief ischemia) is used to assess peripheral microvascular function and is sensitive to the integrated behavior of resistance vessels and local vasodilator pathways [68]. In athlete monitoring, this matters because microvascular function influences substrate delivery and byproduct clearance, which are foundational constraints on the recovery trajectory after damaging work [64].

Swelling as a Fluid-Exchange and Lymphatic-Clearance Problem

Muscle swelling after damaging exercise is well described, and mechanistic discussion increasingly distinguishes between vascular leakage/filtration and lymphatic removal as separate controllable processes [20]. Microvascular fluid exchange is governed by Starling forces, and modern revisions emphasize the role of the endothelial barrier (including the glycocalyx) and the importance of lymphatic return in maintaining steady-state fluid balance [69]. Lymphatic pumping depends on intrinsic collecting-vessel contractions and extrinsic mechanical forces, linking movement and tissue pressure fluctuations to edema clearance [70].

Connecting Swelling to Stiffness and DOMS Phenotypes

MRI studies show edema-like signal changes after eccentric protocols that induce DOMS, supporting swelling as a measurable tissue feature during the soreness window [71]. Quantitative MRI and shear-wave elastography demonstrate that DOMS evolves over days with concurrent changes in tissue water-sensitive measures and mechanical stiffness-related indices [47]. These imaging findings support a practical model where perceived "tightness" can reflect fluid shifts and altered passive mechanical properties alongside nociceptive processes [47].

Practical Box

When circumference, ultrasound appearance, or perceived tightness rises after eccentric stress, the mechanism may include increased filtration/permeability and delayed lymphatic clearance rather than only inflammation. Active recovery plausibly influences this system because muscle contractions can support lymph movement through extrinsic pumping while also modifying perfusion [70].

Highlights

- *Exercise hyperemia is mediated by interacting signals (NO, prostaglandins, ATP) and is protected during work by functional sympatholysis.*
- *Post-exercise swelling is best framed as an imbalance between microvascular fluid formation and lymphatic clearance, not a single-cause phenomenon.*
- *Imaging studies in DOMS support that edema-like changes and stiffness-related measures evolve together across days, helping explain "tightness" as a biomechanical–fluid phenotype.*

1.4.1 From Capillary Filtration to Interstitial Pressure to Lymphatic Clearance

Edema reflects an imbalance between microvascular filtration into the interstitium and lymphatic removal back to the circulation [72]. The interstitial space is mechanically and hydraulically regulated, meaning changes in interstitial volume and pressure can alter tissue mechanics and transport [73]. Lymphatic pumping and permeability mechanisms contribute to edema clearance, which is relevant because post-exercise swelling can influence stiffness sensations and range-of-motion changes [72].

Practical Box

Swelling can elevate passive resistance to movement by altering tissue pressure and geometry, which athletes often describe as tightness [74]. This helps explain why perceived stiffness can rise even when neural drive is unchanged [75].

Highlights

- *Fluid balance is a coupled microvascular–interstitial–lymphatic system.*

- *Interstitial mechanics can change transport and resistance to movement.*
- *Swelling can confound stiffness-based interpretations in athlete monitoring.*

1.4.2 Pressure Gradients in the Microcirculation and Their Relevance to Post-exercise Edema

Fluid exchange within microcirculation is governed by the interaction of hydrostatic and oncotic pressures across the capillary wall. The classical Starling framework describes net fluid movement as the balance between capillary hydrostatic pressure, interstitial hydrostatic pressure, and the difference in colloid osmotic pressures between plasma and interstitial fluid [69, 76]. Capillary hydrostatic pressure (Pc) is highest at the arterial end of the capillary, typically around 30–35 mmHg, and declines along the capillary to approximately 10–15 mmHg at the venous end due to vascular resistance [69, 76]. Elevated Pc promotes filtration of fluid from the intravascular space into the interstitium. In contrast, plasma oncotic pressure (πc), largely generated by albumin, averages approximately 25 mmHg and opposes filtration by favoring fluid reabsorption into the capillary [76]. On the interstitial side, interstitial hydrostatic pressure (Pi) is normally near zero, ranging approximately from -2 to $+2$ mmHg under physiological conditions [72]. Interstitial oncotic pressure (πi) reflects the presence of proteins within the extracellular space and increases when endothelial permeability rises. Mechanical stress, eccentric loading, and inflammatory signaling can increase endothelial permeability, allowing greater protein leakage into the interstitium, elevating πi and thereby promoting sustained net filtration [72].

Contemporary revisions of the Starling principle emphasize the role of the endothelial glycocalyx in regulating effective oncotic gradients and transcapillary filtration. Disruption of the glycocalyx alters the balance between Pc, πc, and πi and may reduce the extent of true reabsorption at the venous end of the capillary, thereby increasing reliance on lymphatic removal of filtered fluid [76]. In this revised model, capillary exchange is not simply a switch from filtration to reabsorption along the capillary length but rather a dynamic system in which a substantial portion of filtered fluid is physiologically cleared by the lymphatic system.

The lymphatic system operates under substantially lower pressures than the blood microcirculation. Initial lymphatic vessels typically function at pressures near 0–2 mmHg, and lymph propulsion depends on pressure gradients between the interstitial space and lymphatic lumen, intrinsic rhythmic contractions of collecting lymphatics, and extrinsic mechanical forces such as skeletal muscle contraction [70]. Consequently, relatively small increases in interstitial hydrostatic pressure can significantly influence lymphatic filling and flow. The pressure differences between the arterial capillary segment (higher Pc), the venous capillary segment (lower Pc with relatively greater oncotic influence), and the low-pressure lymphatic system are central to the dynamics of post-exercise edema. During exercise-induced hyperemia, capillary hydrostatic pressure increases, and in the presence of transiently elevated endothelial permeability, net filtration may exceed lymphatic clearance capacity. If mechanical support of lymph flow (e.g., rhythmic muscle contraction) is reduced during recovery, excess interstitial fluid accumulates [70, 72].

An increase in interstitial fluid volume elevates Pi and alters tissue mechanical properties, increasing passive resistance to movement and contributing to the subjective perception of "stiffness." Importantly, these mechanical changes may occur independently of neural activation changes, explaining why perceived tightness can increase without measurable alterations in motor drive [74, 75].

Highlights

- *Post-exercise edema is not solely a consequence of inflammatory signaling but reflects a shift in the balance between capillary hydrostatic pressure and the capacity of the lymphatic system to maintain low interstitial pressure.*
- *Even modest alterations in pressure gradients between arterial, venous, and lymphatic segments can bias the system toward sustained net filtration.*
- *This mechanistic perspective has direct implications for recovery strategies, as interventions such as active recovery, compression, or limb elevation may function primarily by modifying hydrostatic gradients and enhancing lymphatic drainage rather than directly altering inflammatory pathways.*

1.4.3 Measuring Microvascular Function in Athletes

Near-infrared spectroscopy (NIRS) combined with brief arterial occlusion allows quantification of microvascular oxygenation dynamics and post-occlusive reactivity in skeletal muscle [77]. By tracking changes in deoxygenated and oxygenated hemoglobin during occlusion and reperfusion, NIRS provides insight into tissue oxygen extraction, reperfusion kinetics, and microvascular responsiveness. Because reactive hyperemia outcomes are highly sensitive to protocol parameters, including occlusion duration, cuff pressure, limb position, temperature, hydration status, and sympathetic tone, strict standardization is essential when NIRS is used for longitudinal athlete monitoring [78]. Consequently, measurement selection must always align with the underlying mechanism of interest and the performance or recovery decision that follows [1].

In addition to NIRS, laser Doppler flowmetry (LDF) remains one of the most established and widely validated techniques for assessing cutaneous microcirculation. LDF quantifies relative blood flow by analyzing frequency shifts in reflected laser light caused by moving erythrocytes. It provides continuous, high-temporal-resolution assessment of perfusion and is considered a reference method for evaluating post-occlusive reactive hyperemia [79, 80]. Typical parameters derived from

LDF during occlusion–reperfusion protocols include resting flow (RF), biological zero (BZ), peak hyperemia (Rhmax), time to peak (TP), and recovery time (TR).

Recent work [81] demonstrated that increasing arterial occlusion pressure (AOP) significantly alters microvascular parameters measured with LDF, including Rhmax, TP, and BZ, as well as autonomic indices derived from heart rate variability. Importantly, responses plateaued between 100 and 130% AOP, suggesting a physiological ceiling in microvascular and autonomic modulation. These findings indicate that LDF-based post-occlusion testing may capture both local endothelial responsiveness and early alterations in cardiovascular regulatory dynamics.

A closely related and widely used functional assessment is the post-occlusion reactive hyperemia (PORH) test. This protocol involves transient arterial occlusion (typically five minutes) followed by rapid cuff release and assessment of reperfusion dynamics [79]. The magnitude and kinetics of the hyperemic response reflect the integrated action of endothelial, metabolic, and myogenic mechanisms [82]. While the immediate PORH response is largely driven by local vascular mechanisms, its amplitude and recovery profile may be modulated by systemic autonomic tone, particularly under conditions of repeated occlusive stress, such as blood flow restriction (BFR) training [81]. Therefore, PORH testing may serve not only as an index of microvascular reactivity but also as a sensitive marker of early vascular or autonomic dysregulation.

Beyond LDF and NIRS, additional techniques exist for assessing microvascular function. Laser speckle contrast imaging (LSCI) enables spatial mapping of superficial perfusion across larger tissue areas, although its use remains primarily research based. Venous occlusion plethysmography provides quantitative estimates of limb blood flow by analyzing volume changes during controlled venous occlusion, offering insight into perfusion at the limb level rather than exclusively within superficial microvascular beds. Video capillaroscopy allows direct visualization of capillary density and morphology and is particularly useful in clinical microangiopathy research.

Importantly, the majority of these techniques-including NIRS, LDF, LSCI, capillaroscopy, and PORH protocols-are non-invasive. They do not require tissue penetration, contrast administration, or pharmacological provocation. This non-invasive nature is particularly important in athletic populations, where repeated assessments must be safe, repeatable, and minimally disruptive to training schedules. The ability to perform serial measurements without physiological burden makes microvascular monitoring a feasible addition to integrated athlete readiness frameworks.

Ultimately, each method captures a slightly different physiological dimension. NIRS primarily reflects oxygenation dynamics, LDF and LSCI assess relative perfusion, plethysmography estimates volumetric blood flow, and PORH protocols evaluate vascular reactivity. The interpretive value of any single metric increases substantially when contextualized alongside soreness, functional performance measures, autonomic markers, and recent training exposure.

Practical Box

A club or performance laboratory can implement a standardized 5-min occlusion–reperfusion protocol using NIRS or LDF on a weekly basis to detect meaningful within-athlete changes over time. Absolute values should not be compared between athletes; instead, individual trends should guide interpretation. Decision-making value increases when vascular measures are integrated with soreness ratings, functional outputs (e.g., jump or sprint performance), heart rate variability, and recent training load history.

Highlights

- *NIRS and LDF provide complementary windows into microvascular oxygenation and perfusion dynamics.*
- *Post-occlusion reactive hyperemia (PORH) reflects endothelial and metabolic vasodilatory capacity.*
- *Microvascular responses to arterial occlusion are pressure-dependent but demonstrate physiological saturation beyond certain thresholds.*
- *Most microvascular assessment tools used in sport science are non-invasive and suitable for repeated monitoring.*
- *Interpretation must be individualized and context-driven rather than based on isolated absolute values.*

Near-infrared spectroscopy combined with brief occlusion can quantify aspects of microvascular oxygenation dynamics and post-occlusive reactivity in skeletal muscle [77]. However, reactive hyperemia outcomes are sensitive to protocol and confounders, which means standardization is essential if the goal is longitudinal athlete monitoring [78]. Thus, measurement choices must match the mechanism of interest and the decision that follows [1].

Practical Box

A club can track a standardized occlusion–reperfusion test weekly to detect meaningful within-athlete shifts rather than compare raw values between athletes [78]. The decision value increases when the vascular measure is interpreted alongside soreness, function, and training exposure history [1].

Highlights

- *Near-infrared spectroscopy and occlusion provides a feasible window into microvascular dynamics.*
- *Reactive hyperemia is protocol-sensitive and requires strict standardization.*
- *The interpretation must be connected to the athlete-specific decision context.*

1.5 DOMS and Pain Modulation

Delayed-onset muscle soreness (DOMS) is a common post-exercise pain phenomenon that typically follows novel or eccentric-biased loading and can impair function and training quality [27]. DOMS is linked to mechanical and inflammatory sensitization of nociceptors, including contributions from bradykinin, prostaglandins, cytokines, and nerve growth factor signaling [83]. Because pain can alter motor output and coordination, DOMS is a performance-relevant phenomenon even when gross physiological markers appear recovered [84].

Highlights

- *DOMS is common after novelty and eccentric exposure and can impair performance.*
- *Its mechanisms include nociceptor sensitization by inflammatory mediators.*
- *Pain can change movement and readiness independent of other systems.*

1.5.1 DOMS Is Not a One-to-One Proxy for Damage Magnitude

Soreness can occur without proportional changes in creatine kinase or histological indices, indicating that DOMS is an imperfect proxy for structural disruption [85]. Conversely, meaningful force deficits can persist even after soreness declines, because strength loss can reflect excitation–contraction coupling and remodeling dynamics rather than pain intensity alone [86]. This mismatch is a central justification for multi-domain monitoring rather than soreness-only decision rules [1].

Practical Box

An athlete may report minimal soreness yet still demonstrate suppressed force or power if excitation–contraction coupling recovery lags behind pain resolution [86]. Alternatively, high soreness can reflect sensitization processes even when performance impact is modest, especially in trained athletes with protective adaptations [48].

Highlights

- *DOMS does not map linearly onto damage severity.*
- *Strength and soreness can decouple because they arise from different mechanisms.*
- *Decision rules should combine symptoms with function and exposure history.*

1.5.2 Where DOMS Likely "Lives" Biologically

Evidence supports that connective tissue and fascia may contribute to post-exercise pain generation and stiffness sensations, alongside muscle fibers [87]. Sensitized group III/IV afferents transmit muscle pain and can influence reflexive motor control and central perception of effort [84]. This neurobiology explains why analgesia can transiently improve performance without necessarily accelerating structural repair [84].

Practical Box

Light movement can transiently reduce perceived soreness via short-lived analgesic mechanisms, even though underlying remodeling timelines remain unchanged [27]. This effect is helpful for training management but should not be misread as full biological recovery [2].

Highlights

- *DOMS likely reflects multi-tissue contributions, including connective tissue.*
- *Nociceptive afferents can shape motor output and effort perception.*

- *Symptom relief and tissue repair are not identical endpoints.*

1.6 Tone, Stiffness, and "Tightness"

Muscle tone is often described as resistance to passive stretch, but physiologically it reflects interacting neural circuits and passive viscoelastic tissue properties [75]. Stiffness is a mechanical construct describing the relationship between force and length change, and it can arise from passive elements (titin, ECM) and active cross-bridge–driven behavior [75]. In practice, athletes' reports of "tightness" can reflect increased passive resistance from edema or ECM state, increased reflex-mediated activation, or both [72].

Highlights

- *Tone includes neural and passive viscoelastic components.*
- *Stiffness can be passive (titin/ECM) or active (activation-dependent).*
- *"Tightness" is not a diagnosis, because multiple mechanisms can generate it.*

1.6.1 Passive Mechanisms: Titin, ECM, and Fluid State

Titin phosphorylation and structural state contribute to passive stiffness regulation, demonstrating that stiffness is biologically modifiable rather than fixed [88]. The skeletal muscle ECM provides mechanical scaffolding and transmits force, and its remodeling can change passive mechanical behavior [39]. Edema can modify tissue pressure and geometry, changing passive resistance and thus influencing stiffness sensations even without increased activation [74].

Practical Box

Passive stiffness can rise when interstitial fluid volume increases, because tissue pressure–volume relationships alter mechanical behavior [74]. This helps explain athlete-reported tightness after long travel or heat exposure where fluid shifts are plausible contributors [73].

Highlights

- *Passive stiffness is shaped by titin and ECM biology.*
- *ECM remodeling can change passive force transmission and stiffness.*
- *Fluid-state changes can confound stiffness sensations and measurements.*

1.6.2 Active and Reflex Mechanisms: Spindle Control, Guarding, and Readiness

Tone regulation involves spinal and supraspinal mechanisms and muscle spindle pathways, which can change under pain or threat states [75]. Pain-related guarding can increase background activation and apparent stiffness, which can reduce range of motion and alter movement strategies [84]. Therefore, observing increased tone after DOMS does not automatically imply beneficial readiness, because it may reflect protective motor behavior [27].

Practical Box

Athletes with DOMS often self-select lower joint excursions and different landing mechanics, consistent with pain-modulated motor control [84]. This can reduce training quality or shift load to other tissues, linking soreness biology to injury-relevant mechanics [27].

Highlights

- *Tone can increase via neural regulation, not only via passive tissue state.*
- *Pain can induce guarding that changes mechanics and perceived readiness.*
- *Higher tone can be maladaptive if it reflects threat-protective behavior.*

1.6.3 What Can Bias "Tone" and "Stiffness" Readings

Myotonometry can provide reliable estimates of viscoelastic properties in certain contexts, but validity and comparability depend on population, site, and protocol consistency [89]. Shear-wave elastography is interesting for muscle stiffness assessment, yet outputs are sensitive to joint position, contraction state, and methodological constraints [90]. Therefore, measurement standardization (positioning, timing, muscle state, and repeated measures strategy) is a non-negotiable prerequisite for longitudinal inference in athletes.

Practical Box

Changing hip or knee position alters measured shear modulus in hamstrings, so stiffness change can be positional artifact if angles are not fixed [90]. For DOMS-related applications, elastography measures can change over time alongside MRI indices, reinforcing that timing and standardization determine interpretability [47].

Highlights

- *Reliability does not guarantee validity across all muscles and contexts.*
- *Joint position and contraction state can materially change elastography outcomes.*
- *Without strict protocol control, mechanical biomarkers can mislead decisions.*

1.7 Translating Mechanisms into Monitoring Choices

Recovery monitoring should be linked to the decision it supports (e.g., repeat high-intensity exposure vs. technical session vs. competition), because each decision implies different limiting systems [1]. Time-course logic is essential, because markers reflect different recovery layers and peak at different times after load [2]. A monitoring set typically benefits from combining at least one performance/function measure, one symptom/perception measure, and one physiology layer (autonomic or microvascular) rather than relying on a single surrogate [1].

Practical Box

A team can pair a jump or sprint measure (function), a soreness/tightness rating (perception), and an HRV/HRR or occlusion–reperfusion measure (physiology) to triangulate readiness. This structure reduces the risk of missing layer-specific nonrecovery (e.g., autonomic delay with normal legs, or vice versa).

Highlights

- *Monitoring must be decision-driven, not data-driven.*
- *Time-course mismatches explain many monitoring false positives/negatives.*
- *Triangulation across layers is more biologically valid than single-marker reliance.*

References

1. Kellmann M et al (2018) Recovery and performance in sport: consensus statement. Int J Sports Physiol Perform 13:240–245
2. Skorski S et al (2019) The temporal relationship between exercise, recovery processes, and changes in performance. Int J Sports Physiol Perform 14:1015–1021
3. Bogdanis GC, Nevill ME, Boobis LH, Lakomy HK, Nevill AM (1995) Recovery of power output and muscle metabolites following 30 s of maximal sprint cycling in man. J Physiol 482:467–480
4. Harris RC et al (1976) The time course of phosphorylcreatine resynthesis during recovery of the quadriceps muscle in man. Pflügers Arch Eur J Physiol 367:137–142
5. McMahon S, Jenkins D (2002) Factors affecting the rate of phosphocreatine resynthesis following intense exercise. Sports Med 32:761–784
6. Fitts RH (2008) The cross-bridge cycle and skeletal muscle fatigue. J Appl Physiol 104:551–558
7. Allen DG, Trajanovska S (2012) The multiple roles of phosphate in muscle fatigue. Front Physiol 3
8. Proske U, Morgan DL (2001) Muscle damage from eccentric exercise: mechanism, mechanical signs, adaptation and clinical applications. J Physiol 537:333–345
9. Burke LM, van Loon LJC, Hawley JA (2017) Postexercise muscle glycogen resynthesis in humans. J Appl Physiol 122:1055–1067
10. van Hall G, Shirreffs SM, Calbet JAL (2000) Muscle glycogen resynthesis during recovery from cycle exercise: no effect of additional protein ingestion. J Appl Physiol 88:1631–1636
11. Alghannam A, Gonzalez J, Betts J (2018) Restoration of muscle glycogen and functional capacity: role of post-exercise carbohydrate and protein co-ingestion. Nutrients 10:253
12. Kaikkonen P, Rusko H, Martinmäki K (2008) Post-exercise heart rate variability of endurance athletes after different high-intensity exercise interventions. Scand J Med Sci Sports 18:511–519
13. Addleman JS, Lackey NS, DeBlauw JA, Hajduczok AG (2024) Heart rate variability applications in strength and conditioning: a narrative review. J Funct Morphol Kinesiol 9:93
14. Seiler S, Haugen O, Kuffel E (2007) Autonomic recovery after exercise in trained athletes. Med Sci Sports Exerc 39:1366–1373
15. Daly W, Seegers C, Timmerman S, Hackney AC (2004) Peak cortisol response to exhausting exercise: effect of blood sampling schedule. Med Sport 8:17–20
16. Urhausen A, Gabriel H, Kindermann W (1995) Blood hormones as markers of training stress and overtraining. Sports Med 20:251–276
17. Doherty R, Madigan SM, Nevill A, Warrington G, Ellis JG (2021) The sleep and recovery practices of athletes. Nutrients 13:1330
18. Botonis PG, Grammenou M, Toubekis AG (2025) Athletes and sleep issues: new insights into translating laboratory findings in a real-world setting. Sports Health Multidiscip Approach 17:435–437

19. Charest J, Grandner MA (2020) Sleep and athletic performance. Sleep Med Clin 15:41–57
20. Peake JM, Neubauer O, Della Gatta PA, Nosaka K (2017) Muscle damage and inflammation during recovery from exercise. J Appl Physiol 122:559–570
21. Hody S, Croisier J-L, Bury T, Rogister B, Leprince P (2019) Eccentric muscle contractions: risks and benefits. Front Physiol 10
22. Takekura H, Fujinami N, Nishizawa T, Ogasawara H, Kasuga N (2001) Eccentric exercise-induced morphological changes in the membrane systems involved in excitation—contraction coupling in rat skeletal muscle. J Physiol 533:571–583
23. Allen DG (2001) Eccentric muscle damage: mechanisms of early reduction of force. Acta Physiol Scand 171:311–319
24. Docherty S et al (2022) The effect of exercise on cytokines: implications for musculoskeletal health: a narrative review. BMC Sports Sci Med Rehabil 14:5
25. Hyldahl RD, Olson T, Welling T, Groscost L, Parcell AC (2014) Satellite cell activity is differentially affected by contraction mode in human muscle following a work-matched bout of exercise. Front Physiol 5
26. Vickers AJ (2001) Time course of muscle soreness following different types of exercise. BMC Musculoskelet Disord 2:5
27. Cheung K, Hume PA, Maxwell L (2003) Delayed onset muscle soreness. Sports Med 33:145–164
28. Miller BF et al (2005) Coordinated collagen and muscle protein synthesis in human patella tendon and quadriceps muscle after exercise. J Physiol 567:1021–1033
29. Phillips SM, Tipton KD, Aarsland A, Wolf SE, Wolfe RR (1997) Mixed muscle protein synthesis and breakdown after resistance exercise in humans. Am J Physiol-Endocrinol Metab 273:E99–E107
30. Pilegaard H, Saltin B, Neufer PD (2003) Exercise induces transient transcriptional activation of the PGC-1α gene in human skeletal muscle. J Physiol 546:851–858
31. Stepto NK et al (2012) Short-term intensified cycle training alters acute and chronic responses of PGC1α and cytochrome C oxidase IV to exercise in human skeletal muscle. PLoS One 7:e53080
32. Owens DJ, Twist C, Cobley JN, Howatson G, Close GL (2019) Exercise-induced muscle damage: what is it, what causes it and what are the nutritional solutions? Eur J Sport Sci 19:71–85
33. Morgan DL, Proske U (2004) Popping sarcomere hypothesis explains stretch-induced muscle damage. Clin Exp Pharmacol Physiol 31:541–545
34. Whitehead NP, Weerakkody NS, Gregory JE, Morgan DL, Proske U (2001) Changes in passive tension of muscle in humans and animals after eccentric exercise. J Physiol 533:593–604
35. Warren GL et al (1993) Excitation failure in eccentric contraction-induced injury of mouse soleus muscle. J Physiol 468:487–499
36. Corona BT et al (2010) Junctophilin damage contributes to early strength deficits and EC coupling failure after eccentric contractions. Am J Phys Cell Phys 298:C365–C376
37. Stožer A, Vodopivc P, Križančić Bombek L (2020) Pathophysiology of exercise-induced muscle damage and its structural, functional, metabolic, and clinical consequences. Physiol Res 565–598. https://doi.org/10.33549/physiolres.934371
38. Baird MF, Graham SM, Baker JS, Bickerstaff GF (2012) Creatine-kinase- and exercise-related muscle damage implications for muscle performance and recovery. J Nutr Metab 2012:1–13
39. Gillies AR, Lieber RL (2011) Structure and function of the skeletal muscle extracellular matrix. Muscle Nerve 44:318–331
40. Peter AK, Cheng H, Ross RS, Knowlton KU, Chen J (2011) The costamere bridges sarcomeres to the sarcolemma in striated muscle. Prog Pediatr Cardiol 31:83–88
41. Matsumura K et al (1993) The role of the dystrophin-glycoprotein complex in the molecular pathogenesis of muscular dystrophies. Neuromuscul Disord 3:533–535
42. Gao QQ, McNally EM (2015) The dystrophin complex: structure, function, and implications for therapy. Compr Physiol 1223–1239. https://doi.org/10.1002/cphy.c140048

43. Ramaswamy KS et al (2011) Lateral transmission of force is impaired in skeletal muscles of dystrophic mice and very old rats. J Physiol 589:1195–1208
44. Boppart MD, Mahmassani ZS (2019) Integrin signaling: linking mechanical stimulation to skeletal muscle hypertrophy. Am J Phys Cell Phys 317:C629–C641
45. Herzog W (2018) The multiple roles of titin in muscle contraction and force production. Biophys Rev 10:1187–1199
46. Paulin D, Li Z (2004) Desmin: a major intermediate filament protein essential for the structural integrity and function of muscle. Exp Cell Res 301:1–7
47. Agten CA et al (2017) Delayed-onset muscle soreness: temporal assessment with quantitative MRI and shear-wave ultrasound elastography. Am J Roentgenol 208:402–412
48. McHugh MP (2003) Recent advances in the understanding of the repeated bout effect: the protective effect against muscle damage from a single bout of eccentric exercise. Scand J Med Sci Sports 13:88–97
49. Tidball JG (2017) Regulation of muscle growth and regeneration by the immune system. Nat Rev Immunol 17:165–178
50. Malm C et al (2000) Immunological changes in human skeletal muscle and blood after eccentric exercise and multiple biopsies. J Physiol 529:243–262
51. Butterfield TA, Best TM, Merrick MA (2006) The dual roles of neutrophils and macrophages in inflammation: a critical balance between tissue damage and repair. J Athl Train 41:457–465
52. St. Pierre Schneider B, Tiidus PM (2007) Neutrophil infiltration in exercise-injured skeletal muscle. Sports Med 37:837–856
53. Kawanishi N, Mizokami T, Niihara H, Yada K, Suzuki K (2016) Neutrophil depletion attenuates muscle injury after exhaustive exercise. Med Sci Sports Exerc 48:1917–1924
54. Wang X, Zhou L (2022) The many roles of macrophages in skeletal muscle injury and repair. Front Cell Dev Biol 10
55. Tidball JG (2011) Mechanisms of muscle injury, repair, and regeneration. Compr Physiol 2029–2062. https://doi.org/10.1002/cphy.c100092
56. Tu H, Li Y-L (2023) Inflammation balance in skeletal muscle damage and repair. Front Immunol 14
57. Hindi SM, Kumar A (2016) Toll-like receptor signalling in regenerative myogenesis: friend and foe. J Pathol 239:125–128
58. De Mori R et al (2007) Multiple effects of high mobility group box protein 1 in skeletal muscle regeneration. Arterioscler Thromb Vasc Biol 27:2377–2383
59. Paiva-Oliveira EL et al (2012) TLR4 signaling protects from excessive muscular damage induced by *Bothrops jararacussu* snake venom. Toxicon 60:1396–1403
60. Sass F et al (2018) Immunology guides skeletal muscle regeneration. Int J Mol Sci 19:835
61. Forcina L, Cosentino M, Musarò A (2020) Mechanisms regulating muscle regeneration: insights into the interrelated and time-dependent phases of tissue healing. Cells 9:1297
62. Basil MC, Levy BD (2016) Specialized pro-resolving mediators: endogenous regulators of infection and inflammation. Nat Rev Immunol 16:51–67
63. Mackey AL (2013) Does an NSAID a day keep satellite cells at bay? J Appl Physiol 115:900–908
64. Joyner MJ, Casey DP (2015) Regulation of increased blood flow (hyperemia) to muscles during exercise: a hierarchy of competing physiological needs. Physiol Rev 95:549–601
65. Boushel R et al (2002) Combined inhibition of nitric oxide and prostaglandins reduces human skeletal muscle blood flow during exercise. J Physiol 543:691–698
66. Crecelius AR et al (2015) Contracting human skeletal muscle maintains the ability to blunt α 1-adrenergic vasoconstriction during K IR channel and Na^+/K^+-ATPase inhibition. J Physiol 593:2735–2751
67. Hearon CM et al (2017) Sympatholytic effect of intravascular ATP is independent of nitric oxide, prostaglandins, Na^+/K^+-ATPase and K IR channels in humans. J Physiol 595:5175–5190
68. Rosenberry R, Nelson MD (2020) Reactive hyperemia: a review of methods, mechanisms, and considerations. Am J Phys Regul Integr Comp Phys 318:R605–R618

69. Levick JR (2004) Revision of the Starling principle: new views of tissue fluid balance. J Physiol 557:704
70. Scallan JP, Zawieja SD, Castorena-Gonzalez JA, Davis MJ (2016) Lymphatic pumping: mechanics, mechanisms and malfunction. J Physiol 594:5749–5768
71. Evans GF, Haller RG, Wyrick PS, Parkey RW, Fleckenstein JL (1998) Submaximal delayed-onset muscle soreness: correlations between MR imaging findings and clinical measures. Radiology 208:815–820
72. Breslin JW (2023) Edema and lymphatic clearance: molecular mechanisms and ongoing challenges. Clin Sci 137:1451–1476
73. Stewart RH (2020) A modern view of the interstitial space in health and disease. Front Vet Sci 7
74. Reed RK, Rubin K (2010) Transcapillary exchange: role and importance of the interstitial fluid pressure and the extracellular matrix. Cardiovasc Res 87:211–217
75. Ganguly J, Kulshreshtha D, Almotiri M, Jog M (2021) Muscle tone physiology and abnormalities. Toxins (Basel) 13:282
76. Levick JR, Michel CC (2010) Microvascular fluid exchange and the revised Starling principle. Cardiovasc Res 87:198–210
77. Hendrick E, Jamieson A, Chiesa ST, Hughes AD, Jones S (2024) A short review of application of near-infrared spectroscopy (NIRS) for the assessment of microvascular post-occlusive reactive hyperaemia (PORH) in skeletal muscle. Front Physiol 15
78. Coccarelli A, Nelson MD (2023) Modeling reactive hyperemia to better understand and assess microvascular function: a review of techniques. Ann Biomed Eng 51:479–492
79. Morales F et al (2005) How to assess post-occlusive reactive hyperaemia by means of laser Doppler perfusion monitoring: application of a standardised protocol to patients with peripheral arterial obstructive disease. Microvasc Res 69:17–23
80. Gemae MR et al (2021) Myths and methodologies: reliability of forearm cutaneous vasodilatation measured using laser-Doppler flowmetry during whole-body passive heating. Exp Physiol 106:634–652
81. Trybulski R et al (2025) A controlled comparative study on the effect of arterial occlusion pressure on immediate sympathetic and hyperemic responses. Sci Rep 15:41291
82. Horn AG et al (2022) Post-occlusive reactive hyperemia and skeletal muscle capillary hemodynamics. Microvasc Res 140:104283
83. Murase S et al (2010) Bradykinin and nerve growth factor play pivotal roles in muscular mechanical hyperalgesia after exercise (delayed-onset muscle soreness). J Neurosci 30:3752–3761
84. Mizumura K, Taguchi T (2024) Neurochemical mechanism of muscular pain: insight from the study on delayed onset muscle soreness. J Physiol Sci 74:4
85. Nosaka K, Newton M, Sacco P (2002) Delayed-onset muscle soreness does not reflect the magnitude of eccentric exercise-induced muscle damage. Scand J Med Sci Sports 12:337–346
86. Warren GL, Ingalls CP, Lowe DA, Armstrong RB (2001) Excitation-contraction uncoupling: major role in contraction-induced muscle injury. Exerc Sport Sci Rev 29:82–87
87. MacIntyre DL, Reid WD, McKenzie DC (1995) Delayed muscle soreness. Sports Med 20:24–40
88. Muller AE et al (2014) Acute exercise modifies titin phosphorylation and increases cardiac myofilament stiffness. Front Physiol 5
89. Garcia-Bernal M-I, Heredia-Rizo AM, Gonzalez-Garcia P, Cortés-Vega M-D, Casuso-Holgado MJ (2021) Validity and reliability of myotonometry for assessing muscle viscoelastic properties in patients with stroke: a systematic review and meta-analysis. Sci Rep 11:5062
90. Berrigan WA, Wickstrom J, Farrell M, Alter K (2020) Hip position influences shear wave elastography measurements of the hamstring muscles in healthy subjects. J Biomech 109:109930

Chapter 2
Microvascular Regulation in Skeletal Muscle Recovery

Abstract This chapter examines how microvascular regulation shapes skeletal muscle recovery after exercise and tissue loading. It explains the physiological mechanisms governing post-exercise perfusion, with emphasis on reactive hyperemia, endothelial function, autonomic control, interstitial fluid balance, edema formation, and metabolite clearance. The chapter also clarifies how these processes interact to restore tissue homeostasis, support substrate delivery, and remove by-products associated with fatigue and muscle damage. Beyond mechanism, it discusses how microvascular responses can be assessed in practice and how common perfusion measurements should be interpreted in light of methodological limitations and biological context. By integrating vascular physiology with recovery science, this chapter provides the reader with a concise framework for understanding why microcirculatory function is central to recovery and how it can inform the rationale, interpretation, and application of recovery interventions in sport and exercise contexts.

2.1 Microvascular Network as the Recovery Interface

Skeletal muscle recovery depends on the microcirculation because it is the immediate interface enabling oxygen, substrate, and metabolite flux between blood and myocytes [1]. Capillary–myocyte exchange capacity and red blood cell transit behaviors meaningfully constrain muscle oxygen and metabolite flux during and after exercise [2].

Arteriolar networks do not just "increase flow," they coordinate where flow goes by regulating capillary perfusion distribution via terminal arterioles [3]. This distributed control matters in recovery because reperfusion must match heterogeneous fiber recruitment and damage patterns created by loading [4].

Endurance training expands the functional microvascular network through peripheral cardiovascular adaptations that increase diffusion area and improve extraction capacity [5]. Exercise training also increases muscle blood flow capacity via

R. Trybulski, *Biomarker-Guided Physical Recovery Interventions*,
SpringerBriefs in Forensic and Medical Bioinformatics,
https://doi.org/10.1007/978-981-92-0631-5_2

structural remodeling (e.g., angiogenesis and remodeling of resistance vessels) that changes perfusion reserve available during recovery [6].

Practical Box

Microvascular control in skeletal muscle is dominated by terminal arterioles coordinating capillary perfusion rather than by conduit arteries alone [3]. Training increases microvascular "infrastructure," which increases perfusion capacity and improves exchange conditions relevant to faster restoration of homeostasis post-exercise [5].

Highlights

- *Recovery is partly a microvascular problem: exchange capacity and perfusion distribution determine how quickly substrates arrive and by-products leave.*
- *Terminal arterioles and capillary perfusion regulation are central targets when interpreting "perfusion biomarkers" in recovery studies.*

2.2 Post-exercise Hyperemia and Reactive Hyperemia

Blood flow to active muscle rises markedly during contractions and remains dynamically regulated after exercise to restore ionic, metabolic, and thermal homeostasis [4]. Local vasodilator signals in muscle include multiple redundant mediators (e.g., K^+, nitric oxide, purines, prostaglandins) that collectively shape post-exercise hyperemia and reperfusion [7].

Reactive hyperemia is a reproducible reperfusion phenomenon driven by the interaction of hemodynamic forces and local vasoactive signaling after transient occlusion [8]. Because microvessels are the dominant site of vascular control, post-occlusive reactive hyperemia has value for probing microvascular behavior beyond conduit artery responses [9].

Reactive hyperemia is commonly used as a vascular function test, but its magnitude can be altered by limb position and hydrostatic pressure, which can confound recovery interpretations [10]. In skeletal muscle, mechanosensitive responses to changes in pressure and flow can themselves elicit reactive dilation, meaning the "stimulus" is partly mechanical rather than purely metabolic [8].

Practical Box

Post-occlusive reactive hyperemia is influenced by limb position, so "more hyperemia" does not automatically mean "better endothelial health" without controlling hydrostatic conditions [10]. In athlete monitoring, repeated occlusion–reperfusion tests can track microvascular responsiveness, but only if posture, cuff location, and timing are standardized.

Highlights

- *Post-exercise reperfusion reflects redundant local mediators and mechanical factors, not a single pathway.*
- *Reactive hyperemia is informative, but it is highly protocol-dependent and sensitive to gravitational/hydrostatic context.*

2.3 Endothelial Control of Reperfusion

Endothelial cells transduce shear and agonist signals into vasoactive outputs that regulate resistance vessel tone and spreading vasodilation across networks [11]. In skeletal muscle vessels, shear stress can stimulate release of nitric oxide and prostaglandins, supporting flow-mediated dilation during reperfusion [12].

Nitric oxide is not always essential for reactive hyperemia in human limbs, but it can contribute modestly to exercise hyperemia depending on context and vascular bed [13]. The role of endothelium-derived nitric oxide in exercise blood flow control has been debated for decades, with evidence supporting variable contributions across muscles and experimental conditions [14].

Endothelial function is not only "NO bioavailability," because the endothelium also conducts hyperpolarization through myoendothelial coupling to coordinate vasodilation along branches [11]. This conducted signaling helps explain how localized metabolic dilation in distal arterioles can recruit upstream conductance pathways during and after exercise [15].

Repeated exercise training upregulates endothelial NO bioactivity and improves endothelium-dependent vasodilatory function over time [16]. Muscle-damaging exercise can transiently impair endothelial function systemically, indicating that recovery biology includes vascular dysfunction as a modifiable component [17].

Practical Box

Passive leg movement evokes a largely NO-mediated hyperemia and is used as a sensitive assessment of endothelium-mediated vascular function, illustrating how "movement without load" can probe endothelial responsiveness [18]. After eccentric injury, massage attenuated impairment of flow-mediated dilation in a randomized trial, supporting the idea that some recovery modalities can modify vascular function signals [17].

Highlights

- *Endothelial control of recovery perfusion includes NO/prostanoids and network-level conduction of hyperpolarization.*
- *Vascular function can be temporarily impaired after muscle damage, so post-exercise perfusion signals may reflect dysfunction rather than "successful recovery."*

2.4 Autonomic Control and Functional Sympatholysis in Recovery

Sympathetic vasoconstrictor outflow constrains muscle blood flow, but contracting muscle can blunt this effect via functional sympatholysis to match perfusion to metabolic demand [19]. Functional sympatholysis is mechanistically attributed to locally released substances that reduce α-adrenergic vasoconstrictor effectiveness in active muscle [19].

Sympatholysis is not only an exercise phenomenon, because it can persist briefly into early recovery, shaping blood pressure regulation and post-exercise perfusion distribution [20]. The magnitude and reliability of sympatholysis measurements can vary, so autonomic conclusions require method-aware interpretation [21].

Endothelium-to-smooth muscle signaling can moderate adrenergic vasoconstriction, linking endothelial health directly to autonomic-perfusion integration [11]. Therefore, a recovery modality that changes endothelial function plausibly changes the autonomic constraint on reperfusion without any change in central sympathetic drive [11].

Practical Box

Functional sympatholysis can remain evident up to 10 min after exercise, which is directly relevant to how you time post-session recovery interventions [20]. Measurement work shows test–retest considerations for sympatholysis quantification, implying that athlete monitoring requires repeated standardized protocols, not one-off tests [21].

Highlights

- *Autonomic control in recovery is not simply sympathetic on/off, because local factors can override vasoconstriction in a time-dependent manner.*
- *Endothelial signaling can directly shape adrenergic responsiveness, creating a mechanistic bridge between endothelial biomarkers and autonomic effects.*

2.5 Edema and Interstitial Fluid Dynamics after Loading

Eccentric or unfamiliar exercise can produce swelling that reflects injury-linked changes in tissue water distribution and inflammatory processes [22]. Imaging and circumference/ultrasound measures show that muscle swelling follows a time course after eccentric exercise consistent with injury and inflammation [23].

MRI T2 signal changes after eccentric exercise can persist and are used as indicators of edema and altered muscle water characteristics across recovery [24]. T2 relaxation time and volume changes have been used to track muscle damage and adaptation patterns after eccentric work, demonstrating that edema signatures are dynamic rather than instantaneous [24].

Local interstitial volume can alter reflex cardiovascular responses from skeletal muscle, indicating that tissue hydration status is not a passive byproduct but a variable interacting with control systems [25]. Repetitive submaximal exercise can also increase T2-derived edema indicators alongside mechanical property changes, supporting a link between fluid shifts and tissue mechanics [26].

Practical Box

In athlete contexts, MRI T2 elevations can outlast soreness, meaning "feels recovered" and "tissue water normalized" can diverge [27]. Compression garments have been tested against intramuscular edema measures, illustrating that not all mechanical interventions reliably reduce imaging-defined edema [28].

Highlights

- *Post-exercise swelling is measurable and time-dependent, and it can be tracked with MRI/ultrasound/circumference proxies.*
- *Interstitial fluid status interacts with physiological control, so edema is both an outcome and a regulator-relevant state variable.*

2.6 Metabolite Clearance and the Lymphatic/Venous "Pumps"

Lymphatic transport is driven by extrinsic forces including skeletal muscle contractions, respiration, and arterial pulsations, which collectively support macromolecular and fluid clearance [29]. Initial lymphatics absorb interstitial fluid with small pressure gradients, linking interstitial pressure fluctuations to lymph formation and clearance capacity [30].

Human studies using scintographic approaches show that dynamic and isometric contractions modify lymph flow dynamics within exercising skeletal muscle [31]. Leg lymph flow and protein output can increase substantially during prolonged muscular exercise, consistent with exercise-enhanced clearance of filtered fluid and proteins [32].

Active recovery clears blood lactate faster than passive recovery in an intensity-dependent manner, supporting the use of low-to-moderate continued movement to accelerate metabolite kinetics after maximal work [33]. Active recovery near the lactate threshold yields high clearance rates, suggesting that recovery perfusion and metabolic utilization pathways work together in lactate removal [34].

Manual lymph drainage applied after exercise has been associated with faster decreases in serum muscle enzyme levels, consistent with altered post-exercise fluid/solute handling [35]. In combat sports athletes, lymphatic drainage methods have been reported to improve postexercise regeneration indicators in the forearm, illustrating a sport-specific application model [36].

Practical Box

Dynamic contractions increase lymph flow in human skeletal muscle, which provides a mechanistic basis for cool-down movement beyond lactate narratives alone [31]. Post-treadmill manual lymph drainage has been linked to faster declines in serum muscle enzymes, aligning with the idea that clearance pathways can be manipulated [35].

Highlights

- *Metabolite and fluid clearance depend on both blood flow regulation and lymphatic/venous pump mechanics.*
- *Active recovery has strong physiological support for faster lactate kinetics compared with passive recovery, with an intensity-dependent optimum.*

2.7 Interpreting Perfusion Measurements in Practice

Perfusion measurement tools vary in what they truly quantify (e.g., microvascular flux, oxygenation, or blood volume proxies), so more perfusion depends on modality-specific assumptions [37]. Near-infrared spectroscopy with a vascular occlusion test can provide noninvasive indices of microvascular function and metabolism with demonstrated reliability under defined conditions [38].

Laser Doppler flowmetry and NIRS have measurable test–retest reliability characteristics during exercise and recovery, which should shape biomarker selection and sample size logic [39]. Anatomical heterogeneity such as perforator vessel distribution can explain variability in NIRS and LDF signals, reinforcing the need for precise probe placement and within-subject standardization [40].

Contrast-enhanced ultrasound can quantify skeletal muscle microvascular blood flow and has documented repeatability at baseline and in response to exercise/recovery paradigms [41]. Arterial spin labeling MRI can noninvasively measure muscle perfusion with known limitations and reliability constraints that depend on protocol design and signal-to-noise properties [42].

Skin perfusion and muscle oxygenation can respond differently to posture and local cooling, so superficial vascular changes can confound muscle-directed interpretations if the method has skin sensitivity [43]. Reactive hyperemia results are affected by limb position and hydrostatic pressure, so perfusion biomarkers must be interpreted relative to standardized body geometry [10].

Practical Box

Contrast-enhanced ultrasound has published repeatability data for measuring skeletal muscle microvascular blood flow across sessions, supporting its use in intervention trials when operators are trained and protocols are fixed [41]. NIRS-VOT methods can be reliable but remain sensitive to placement and physiology, so athlete monitoring should use within-athlete baselines and strict replication [38].

Highlights

- *Different tools measure different physiological constructs, so "perfusion" is not a single biomarker.*
- *Reliability, probe placement, limb position, and skin contamination are dominant confounders in recovery perfusion interpretation.*

2.7.1 Near-Infrared Spectroscopy (NIRS) and Vascular Occlusion Testing

Near-infrared spectroscopy (NIRS) estimates local muscle oxygenation from the differential absorption of near-infrared light by oxygenated and deoxygenated chromophores, and in skeletal muscle the signal includes contributions from hemoglobin and myoglobin [44]. For that reason, NIRS should be interpreted primarily as an index of local oxygenation balance and oxygen extraction behavior, not as a direct volumetric measure of muscle blood flow [44]. In recovery contexts, the most informative NIRS applications are usually continuous monitoring after exercise and vascular occlusion tests that quantify desaturation during ischemia and resaturation or reperfusion kinetics after cuff release [45].

A typical NIRS vascular occlusion procedure places the probe over a predefined muscle, stabilizes the limb, records baseline StO2 or TSI, applies supra-systolic cuff occlusion for a fixed period, and then quantifies the reperfusion upslope over the first seconds after release [45]. The occlusion duration matters because reperfusion slope becomes steeper and usually more discriminative after longer ischemic stimuli, with 5-min protocols generally producing stronger vascular reactivity signals than shorter occlusions [46].

Reliability of NIRS-derived reperfusion slope is often better than ultrasound flow-mediated dilation, but reliability worsens when adipose tissue is thicker, when placement is inconsistent, or when results are stratified by sex without controlling site-specific tissue thickness [45].

Tips

- *Place the optode over a clearly defined muscle belly, document the anatomical site, and replicate probe position across sessions because small placement changes alter the sampled vascular territory.*
- *Measure or at least report adipose tissue thickness at the probe site because overlying adipose tissue can overestimate apparent muscle oxygenation and reduce sensitivity to deoxygenation.*
- *Standardize cuff location, occlusion pressure, occlusion duration, posture, and pre-test rest because the reperfusion signal is protocol-dependent.*
- *Prefer within-subject longitudinal interpretation over universal cutoffs because absolute StO2 values are device-, tissue-, and processing-dependent.*

Representative Values

- *In healthy volunteers undergoing a 3-min thenar ischemia-reperfusion test, the post-occlusive StO2 upslope was approximately 4.7–4.8%/s, whereas septic patients showed markedly slower values around 2.3%/s [47].*
- *In a calf dynamic vascular occlusion test in healthy young adults, resaturation rates around 30%/min were reported, but those values should be treated as protocol-specific examples rather than normative thresholds [48].*
- *In reliability work, NIRS reperfusion slope showed intraday and interday CVs around 9% and 14%, respectively, under tightly standardized conditions [45].*

2.7.2 Laser Doppler and Laser Speckle Methods

Laser Doppler techniques quantify flux from moving red blood cells within the sampled optical field, so they are best considered microcirculatory flux methods expressed in arbitrary perfusion units rather than absolute volumetric blood-flow methods [49]. In musculoskeletal recovery research, this makes laser methods attractive for detecting rapid relative changes in superficial microcirculation, but less appropriate for inferring whole-muscle perfusion unless the measurement geometry is carefully controlled [50].

A standard protocol records a stable baseline at constant temperature and posture, applies an intervention or ischemic challenge, and then analyzes relative changes in perfusion units, time-to-peak, or normalized conductance [39]. The main limitation is that optical laser signals are highly sensitive to skin blood flow, local temperature, probe contact, and limb position, which means cutaneous vasomotion can be mistaken for muscle recovery effects [43]. This problem is particularly relevant after cryotherapy, heating, massage, or compression because these interventions often alter skin circulation substantially even when deeper muscle perfusion changes are smaller or different in direction [51].

Tips

- *Control room temperature, local skin temperature, acclimation time, and body position because laser-derived perfusion changes strongly with these perturbations.*
- *Report whether the method sampled skin, intramuscular tissue, or both, because superficial and deep responses are not interchangeable.*
- *Treat perfusion units as relative, device-specific units and avoid cross-device comparison of absolute values.*
- *When exercise is part of the protocol, allow at least 10 min of steady exercise if the aim is to improve reliability of laser and allied optical measures [39].*

Representative Values

- *In an intramuscular anterior tibialis laser-Doppler study, resting muscle flux was approximately 3.5–4 times higher than the simultaneous skin flux measured at the calf or foot [50].*
- *After 3 min of arterial occlusion in that same model, peak muscle flux occurred at about 18.7 ± 9.8 s, and the relative reactive hyperemia increase in muscle was about 2.7 ± 1.3-fold [50].*
- *In exercise reliability work, laser-Doppler perfusion units showed CVs roughly between 18.7 and 28.4%, while normalized cutaneous conductance was even more variable [39].*

2.7.3 Doppler Ultrasound and Passive Leg Movement Paradigms

Doppler ultrasound estimates conduit artery blood flow from vessel diameter and blood velocity, which makes it a strong method for limb inflow quantification even though it does not directly image capillary exchange [52]. For resting and light-to-moderate exercise conditions, Doppler ultrasound is considered a practical and recommendable noninvasive method when performed by an experienced operator [52]. In recovery studies, its major strength is the ability to quantify rapid limb inflow changes after exercise, ischemia, or passive movement with high temporal resolution [53]. Passive leg movement and single passive leg movement are especially useful adjunct paradigms because they evoke a robust hyperemia that is largely nitric-oxide mediated and therefore provide a lower-limb vascular function test that is more muscle-specific than many brachial assessments [53].

Validity is supported by agreement between Doppler-derived limb blood flow and venous occlusion plethysmography during calf exercise, although the two methods may diverge slightly at high intensities because they do not sample identical vascular regions [54]. Reliability is generally strong within the same day and acceptable between days when the same sonographer is used, but operator skill and image reacquisition remain major determinants of error [55].

Tips

- *Record blood velocity and vessel diameter at the same arterial segment across sessions, and document probe angle, insonation site, and averaging window.*
- *Use the same sonographer whenever possible because between-rater variability is substantially larger than within-day repeated measurement error.*
- *If using PLM or sPLM, standardize joint excursion, movement cadence, posture, and analysis window because peak response and area-under-the-curve are protocol-sensitive [53].*
- *Interpret Doppler flow as limb inflow and not as direct microvascular perfusion, unless it is explicitly linked to a microvascular test or imaging method [52].*

Representative Values

- *In classic knee-extensor work, femoral artery blood flow is about 0.3 L/min at rest and can rise to roughly 6–10 L/min during dynamic exercise [56].*
- *During passive leg movement at 60 rpm in healthy young men, leg blood flow increased from about 0.3 ± 0.1 to 0.9 ± 0.1 L/min at 20 s [57].*
- *In single passive leg movement, healthy men showed a peak leg blood flow change of about 719 ± 238 mL/min under control conditions, with the response falling markedly during nitric oxide synthase inhibition [58].*

- *In submaximal single-leg knee-extensor exercise, Doppler ultrasound showed within-day CVs around 4.0–4.3% and between-day CVs around 10.1–20.2% [55].*

2.7.4 Contrast-Enhanced Ultrasound (CEUS)

CEUS assesses skeletal muscle microvascular perfusion by tracking intravascular microbubbles and fitting the replenishment or bolus transit kinetics of acoustic intensity within a region of interest [59]. Its principal advantage over conduit-artery Doppler is that it interrogates the microvascular bed more directly and can separate blood volume and microvascular flux-rate components from the same dataset [59]. For recovery physiology, this makes CEUS especially valuable when the research question concerns capillary recruitment, post-exercise reperfusion kinetics, or failure of microvascular perfusion despite preserved large-artery inflow [59].

Normal procedures generally use either bolus injection with time-intensity analysis or continuous infusion with destruction-replenishment imaging, and both require careful background subtraction and stable transducer positioning [59]. Validity is supported by significant correlations between CEUS-derived muscle perfusion indices and venous occlusion plethysmography [60]. Repeatability appears good in some exercise-based protocols, but reliability can be poor during reactive hyperemia protocols, so the acquisition paradigm matters as much as the technology itself [41].

Tips

- *Report whether the study used bolus or continuous-infusion CEUS because the kinetic parameters and interpretation differ between these approaches.*
- *Define the region of interest anatomically and keep the probe fixed because small transducer shifts change the sampled muscle and background interfaces.*
- *Distinguish blood-volume parameters from flux-rate parameters in reporting because an intervention may change one component without changing the other.*
- *Prefer exercise- or contraction-based CEUS paradigms when the aim is longitudinal repeatability in healthy muscle, because some post-occlusive protocols show much wider CVs [41].*

Representative Values

- *In healthy volunteers, CEUS of the right biceps showed perfusion values around 3.0 ± 2.3 at rest and 22.9 ± 11.0 after exercise, with blood-flow velocity increasing from 0.41 ± 0.24 to 0.64 ± 0.39 mm/s [60].*
- *In a gastrocnemius real-time CEUS study, resting blood volume averaged 3.48 with a range of 0.60–9.92, and the post-exercise maximum rose to about 8.88 [61].*
- *In a healthy-vastus lateralis repeatability study, Bland-Altman bias for CEUS-derived blood-volume and velocity indices was small across sessions, supporting longitudinal use under standardized conditions [41].*
- *In contrast, a calf reactive-hyperemia CEUS study found overall within-individual CVs ranging from 15 to 87%, with time-to-peak being the most reliable parameter [62].*

2.7.5 Arterial Spin Labeling MRI (ASL-MRI)

ASL-MRI measures tissue perfusion by magnetically labeling inflowing arterial blood water and then quantifying the delivery of that labeled blood into muscle without contrast injection [37]. Its major strength is spatially resolved muscle-specific perfusion mapping, which is particularly useful when recovery questions depend on regional heterogeneity across muscles or across compartments within the same muscle group [37]. Its main limitation is intrinsically low signal, which reduces temporal resolution and makes long acquisition windows or exercise paradigms necessary for robust repeated measurement [42].

A practical ASL protocol usually combines supine exercise or post-exercise imaging inside the scanner with repeated signal averaging over a defined muscle region [63]. Validity is supported by its ability to discriminate healthy from disease-impaired skeletal muscle perfusion and by the broader physiological coherence of ASL-derived perfusion with exercise intensity [63]. For recovery research, ASL is therefore most defensible when the priority is regional resolution and mechanistic depth, rather than bedside convenience or immediate post-session feasibility [37].

Tips

- *Use extended acquisition windows or repeated averages because short sampling intervals markedly worsen reliability.*
- *Standardize exercise modality, workload, and timing of image acquisition relative to contraction cessation because ASL signal is highly time-sensitive.*

- *Report the exact muscle or compartment analyzed because ASL is one of the few methods that can exploit spatial heterogeneity, and that advantage is lost if the region of interest is poorly described.*
- *Reserve ASL for questions where regional muscle specificity justifies the logistical burden, because simpler methods are often adequate for whole-limb inflow or superficial optical monitoring.*

Representative Values

- *In healthy adults, steady-state quadriceps perfusion measured by ASL was about 35.0 ± 5.1 mL/min/100 g during light exercise and 51.3 ± 5.6 mL/min/100 g during moderate exercise when data were averaged over 9 min [42].*
- *In peak-exercise calf imaging at 3T, healthy volunteers showed perfusion around 80 ± 23 mL/min/100 g, whereas peripheral arterial disease patients showed lower values around 49 ± 16 mL/min/100 g [63].*
- *In the same peak-exercise study, repeated ASL measurements showed an intraclass correlation coefficient of 0.87, indicating good reproducibility when the protocol is tightly controlled [63].*
- *By contrast, shortening the quadriceps acquisition window from 9 min to 1 min 48 s increased CVs to approximately 28–33% and sharply reduced ICCs [42].*

2.7.6 Remarks

Perfusion measurements should never be treated as interchangeable because NIRS mainly reflects local oxygenation behavior, laser methods mainly reflect superficial microvascular flux, Doppler ultrasound quantifies conduit inflow, CEUS interrogates microvascular blood volume and flux-rate kinetics, and ASL-MRI provides spatially resolved tissue perfusion [44, 49]. Accordingly, the best method is not the most sophisticated one, but the one whose physiological output matches the mechanistic question being asked in the recovery context [52]. Thus, recovery studies should prioritize standardized protocols, repeated within-subject comparisons, and method-specific interpretation rather than generic claims that an intervention improved perfusion [44].

2.8 Translational Implications for Recovery Interventions

If endothelial function is transiently impaired after muscle-damaging exercise, then modalities that modulate endothelial responsiveness can plausibly alter recovery perfusion and downstream exchange [17]. Massage therapy has been shown to attenuate impairment of flow-mediated dilation after exertion-induced muscle injury, providing an existence proof that recovery techniques can change vascular function markers [17].

Because lymph flow increases with contractions and can increase during prolonged exercise, interventions that restore low-level cyclic mechanical pumping may support edema and macromolecule clearance during recovery [32]. Manual lymph drainage has been associated with faster reductions in serum muscle enzymes after treadmill exercise, supporting its candidacy when clearance-limited recovery is suspected [35].

Since active recovery intensity strongly shapes lactate clearance kinetics, prescribing recovery movement can be treated as a dose variable rather than a generic recommendation [34]. Given the redundancy of vasodilator mediators in active hyperemia, single-mechanism explanations for recovery modalities are usually incomplete, so multimarker panels are more defensible than one perfusion endpoint alone [7].

Practical Box

In practice, we can align the intervention with the suspected bottleneck: endothelial impairment signals may justify modalities shown to affect FMD after muscle injury [17]. When edema or heaviness dominates, drainage-oriented approaches have published associations with faster post-exercise enzyme declines and improved regeneration indicators in athletes [35].

Highlights

- *Translate microvascular biology into decisions by matching the dominant limiter (endothelial function, autonomic constraint, edema/clearance, or measurement artifact) to the modality.*
- *Treat recovery movement as a quantifiable dose because lactate clearance is intensity-dependent rather than binary active versus passive.*

References

1. Poole DC (2019) Edward F. Adolph Distinguished Lecture. Contemporary model of muscle microcirculation: gateway to function and dysfunction. J Appl Physiol 127:1012–1033
2. Poole DC, Copp SW, Ferguson SK, Musch TI (2013) Skeletal muscle capillary function: contemporary observations and novel hypotheses. Exp Physiol 98:1645–1658
3. Segal SS (2005) Regulation of blood flow in the microcirculation. Microcirculation 12:33–45
4. Joyner MJ, Casey DP (2015) Regulation of increased blood flow (hyperemia) to muscles during exercise: a hierarchy of competing physiological needs. Physiol Rev 95:549–601
5. Hellsten Y, Nyberg M (2015) Cardiovascular adaptations to exercise training. Compr Physiol 1–32. https://doi.org/10.1002/cphy.c140080
6. Laughlin MH, Roseguini B (2008) Mechanisms for exercise training-induced increases in skeletal muscle blood flow capacity: differences with interval sprint training versus aerobic endurance training. J Physiol Pharmacol 59:71–88
7. Murrant CL, Sarelius IH (2015) Local control of blood flow during active hyperaemia: what kinds of integration are important? J Physiol 593:4699–4711
8. Koller A, Bagi Z (2002) On the role of mechanosensitive mechanisms eliciting reactive hyperemia. Am J Phys Heart Circ Phys 283:H2250–H2259
9. Horn AG et al (2022) Post-occlusive reactive hyperemia and skeletal muscle capillary hemodynamics. Microvasc Res 140:104283
10. Jasperse JL, Shoemaker JK, Gray EJ, Clifford PS (2015) Positional differences in reactive hyperemia provide insight into initial phase of exercise hyperemia. J Appl Physiol 119:569–575
11. Behringer EJ, Segal SS (2012) Spreading the signal for vasodilatation: implications for skeletal muscle blood flow control and the effects of ageing. J Physiol 590:6277–6284
12. Koller A, Dörnyei G, Kaley G (1998) Flow-induced responses in skeletal muscle venules: modulation by nitric oxide and prostaglandins. Am J Phys Heart Circ Phys 275:H831–H836
13. Joyner MJ, Dietz NM (1997) Nitric oxide and vasodilation in human limbs. J Appl Physiol 83:1785–1796
14. McAllister RM, Hirai T, Musch TI (1995) Contribution of endothelium-derived nitric oxide (EDNO) to the skeletal muscle blood flow response to exercise. Med Sci Sports Exerc 27:1145–1151
15. Sinkler SY, Segal SS (2017) Rapid versus slow ascending vasodilatation: intercellular conduction versus flow-mediated signalling with tetanic versus rhythmic muscle contractions. J Physiol 595:7149–7165
16. Maiorana A, O'Driscoll G, Taylor R, Green D (2003) Exercise and the nitric oxide vasodilator system. Sports Med 33:1013–1035
17. Franklin NC, Ali MM, Robinson AT, Norkeviciute E, Phillips SA (2014) Massage therapy restores peripheral vascular function after exertion. Arch Phys Med Rehabil 95:1127–1134
18. Trinity JD et al (2021) The role of the endothelium in the hyperemic response to passive leg movement: looking beyond nitric oxide. Am J Phys Heart Circ Phys 320:H668–H678
19. Saltin B, Mortensen SP (2012) Inefficient functional sympatholysis is an overlooked cause of malperfusion in contracting skeletal muscle. J Physiol 590:6269–6275
20. Moynes J, Bentley RF, Bravo M, Kellawan JM, Tschakovsky ME (2013) Persistence of functional sympatholysis post-exercise in human skeletal muscle. Front Physiol 4
21. Teixeira AL, Garland M, Lee JB, Nardone M, Millar PJ (2022) Assessing functional sympatholysis during rhythmic handgrip exercise using Doppler ultrasound and near-infrared spectroscopy: sex differences and test-retest reliability. Am J Phys Regul Integr Comp Phys 323:R810–R821
22. Peake JM, Neubauer O, Della Gatta PA, Nosaka K (2017) Muscle damage and inflammation during recovery from exercise. J Appl Physiol 122:559–570
23. Nosaka K, Clarkson P (1996) Changes in indicators of inflammation after eccentric exercise of the elbow flexors. Med Sci Sports Exerc 28:953–961
24. Foley JM, Jayaraman RC, Prior BM, Pivarnik JM, Meyer RA (1999) MR measurements of muscle damage and adaptation after eccentric exercise. J Appl Physiol 87:2311–2318

25. Efeld D, Baum K (1996) Influence of gravity on cardiovascular reflexes from skeletal muscle receptors. Med Sci Sports Exerc 28:23–28
26. Sesto ME, Radwin RG, Block WF, Best TM (2005) Anatomical and mechanical changes following repetitive eccentric exertions. Clin Biomech 20:41–49
27. McCully K, Shellock FG, Bank WJ, Posner JD (1992) The use of nuclear magnetic resonance to evaluate muscle injury. Med Sci Sports Exerc 24:537–542
28. Heiss R et al (2018) Effect of compression garments on the development of edema and soreness in delayed-onset muscle soreness (DOMS). J Sports Sci Med 17:392–401
29. Gashev AA (2002) Physiologic aspects of lymphatic contractile function. Ann N Y Acad Sci 979:178–187
30. Schmid-Schonbein GW (1990) Microlymphatics and lymph flow. Physiol Rev 70:987–1028
31. Havas E et al (1997) Lymph flow dynamics in exercising human skeletal muscle as detected by scintography. J Physiol 504:233–239
32. Olszewski W, Engeset A, Icger PM, Sokolowski J, Theodorsen L (1977) Flow and composition of leg lymph in normal men during venous stasis, muscular activity and local hyperthermia. Acta Physiol Scand 99:149–155
33. Menzies P et al (2010) Blood lactate clearance during active recovery after an intense running bout depends on the intensity of the active recovery. J Sports Sci 28:975–982
34. Devlin J et al (2014) Blood lactate clearance after maximal exercise depends on active recovery intensity. J Sports Med Phys Fitness 54:271–278
35. Schillinger A et al (2006) Effect of manual lymph drainage on the course of serum levels of muscle enzymes after treadmill exercise. Am J Phys Med Rehabil 85:516–520
36. Zebrowska A, Trybulski R, Roczniok R, Marcol W (2019) Effect of physical methods of lymphatic drainage on postexercise recovery of mixed martial arts athletes. Clin J Sport Med 29:49–56
37. Richardson RS, Haseler LJ, Nygren AT, Bluml S, Frank LR (2001) Local perfusion and metabolic demand during exercise: a noninvasive MRI method of assessment. J Appl Physiol 91:1845–1853
38. Rogers EM, Banks NF, Jenkins NDM (2023) Metabolic and microvascular function assessed using near-infrared spectroscopy with vascular occlusion in women: age differences and reliability. Exp Physiol 108:123–134
39. Choo HC et al (2017) Reliability of laser Doppler, near-infrared spectroscopy and Doppler ultrasound for peripheral blood flow measurements during and after exercise in the heat. J Sports Sci 35:1715–1723
40. Binzoni T, Leung T, Delpy DT, Fauci MA, Rüfenacht D (2004) Mapping human skeletal muscle perforator vessels using a quantum well infrared photodetector (QWIP) might explain the variability of NIRS and LDF measurements. Phys Med Biol 49:N165–N173
41. Jones EJ, Atherton PJ, Piasecki M, Phillips BE (2023) Contrast-enhanced ultrasound repeatability for the measurement of skeletal muscle microvascular blood flow. Exp Physiol 108:549–553
42. Fulford J, Vanhatalo A (2016) Reliability of arterial spin labelling measurements of perfusion within the quadriceps during steady-state exercise. Eur J Sport Sci 16:80–87
43. Krite Svanberg E, Wollmer P, Andersson-Engels S, Åkeson J (2011) Physiological influence of basic perturbations assessed by non-invasive optical techniques in humans. Appl Physiol Nutr Metab 36:946–957
44. Barstow TJ (2019) Understanding near infrared spectroscopy and its application to skeletal muscle research. J Appl Physiol 126:1360–1376
45. McLay KM, Nederveen JP, Pogliaghi S, Paterson DH, Murias JM (2016) Repeatability of vascular responsiveness measures derived from near-infrared spectroscopy. Physiol Rep 4:e12772
46. McLay KM, Gilbertson JE, Pogliaghi S, Paterson DH, Murias JM (2016) Vascular responsiveness measured by tissue oxygen saturation reperfusion slope is sensitive to different occlusion durations and training status. Exp Physiol 101:1309–1318

47. Creteur J et al (2007) The prognostic value of muscle StO2 in septic patients. Intensive Care Med 33:1549–1556
48. Zamparini G et al (2015) Noninvasive assessment of peripheral microcirculation by near-infrared spectroscopy: a comparative study in healthy smoking and nonsmoking volunteers. J Clin Monit Comput 29:555–559
49. Carolan-Rees G, Tweddel AC, Naka KK, Griffith TM (2002) Fractal dimensions of laser Doppler flowmetry time series. Med Eng Phys 24:71–76
50. Hoffmann U et al (1995) Simultaneous assessment of muscle and skin blood fluxes with the laser-Doppler technique. Int J Microcirc 15:53–59
51. Versteeg N et al (2024) Short-term cutaneous vasodilatory and thermosensory effects of topical methyl salicylate. Front Physiol 15
52. Gliemann L, Mortensen SP, Hellsten Y (2018) Methods for the determination of skeletal muscle blood flow: development, strengths and limitations. Eur J Appl Physiol 118:1081–1094
53. Trinity JD et al (2012) Nitric oxide and passive limb movement: a new approach to assess vascular function. J Physiol 590:1413–1425
54. Murphy E, Rocha J, Gildea N, Green S, Egaña M (2018) Venous occlusion plethysmography vs. Doppler ultrasound in the assessment of leg blood flow kinetics during different intensities of calf exercise. Eur J Appl Physiol 118:249–260
55. Hartmann JP et al (2023) Doppler ultrasound-based leg blood flow assessment during single-leg knee-extensor exercise in an uncontrolled setting. J Vis Exp. https://doi.org/10.3791/65746
56. Saltin B, Radegran G, Koskolou MD, Roach RC (1998) Skeletal muscle blood flow in humans and its regulation during exercise. Acta Physiol Scand 162:421–436
57. Mortensen SP, Askew CD, Walker M, Nyberg M, Hellsten Y (2012) The hyperaemic response to passive leg movement is dependent on nitric oxide: a new tool to evaluate endothelial nitric oxide function. J Physiol 590:4391–4400
58. Broxterman RM et al (2017) Single passive leg movement assessment of vascular function: contribution of nitric oxide. J Appl Physiol 123:1468–1476
59. Nguyen T, Davidson BP (2019) Contrast enhanced ultrasound perfusion imaging in skeletal muscle. J Cardiovasc Imaging 27:163
60. Krix M et al (2005) Assessment of skeletal muscle perfusion using contrast-enhanced ultrasonography. J Ultrasound Med 24:431–441
61. Krix M, Krakowski-Roosen H, Kauczor H-U, Delorme S, Weber M-A (2009) Real-time contrast-enhanced ultrasound for the assessment of perfusion dynamics in skeletal muscle. Ultrasound Med Biol 35:1587–1595
62. Thomas KN, Cotter JD, Lucas SJE, Hill BG, van Rij AM (2015) Reliability of contrast-enhanced ultrasound for the assessment of muscle perfusion in health and peripheral arterial disease. Ultrasound Med Biol 41:26–34
63. Pollak AW et al (2012) Arterial spin labeling MR imaging reproducibly measures peak-exercise calf muscle perfusion. JACC Cardiovasc Imaging 5:1224–1230

Chapter 3
Neuromechanical Biomarkers

Abstract This chapter examines neuromechanical biomarkers as objective indicators of tissue state, with emphasis on muscle tone, stiffness, and elasticity. It clarifies the physiological meaning of each construct and explains how these properties emerge from the interaction of neural drive, muscle architecture, extracellular matrix organization, fluid balance, and viscoelastic tissue behavior. The chapter also reviews the main technologies used to quantify these parameters, including myotonometry, elastography, tensiomyography, and related biomechanical approaches, highlighting what each method truly measures and where interpretation can fail. Particular attention is given to measurement validity, reliability, and the major confounders that influence results, such as posture, probe placement, contraction state, hydration, temperature, prior exercise, pain, and operator technique. Finally, the chapter translates this evidence into practical guidance for sport and clinical contexts, helping readers interpret neuromechanical signals with greater rigor, caution, and applied relevance.

3.1 Why Neuromechanical Biomarkers Belong in Recovery Science

Objective recovery monitoring benefits from markers that are proximal to local tissue state rather than inferred only from soreness or whole-body performance, and localized mechanical measures are attractive because they can be repeated non-invasively at the bedside, clinic, or trackside [1]. Muscle tone, stiffness, and elasticity are related but non-equivalent constructs, and together they describe how tissue resists deformation, stores and dissipates mechanical energy, and returns toward baseline after loading [2]. Because delayed-onset soreness, edema, altered contraction, and connective-tissue loading can all modify these properties, neuromechanical biomarkers can complement perfusion and pain measures when mapping recovery trajectories [3].

R. Trybulski, *Biomarker-Guided Physical Recovery Interventions*,
SpringerBriefs in Forensic and Medical Bioinformatics,
https://doi.org/10.1007/978-981-92-0631-5_3

Practical Box

Reviews of exercise-induced muscle damage support shear wave elastography as a promising monitoring tool, but they also emphasize protocol dependence and incomplete agreement with soreness or MRI-based outcomes [3]. In athlete contexts, localized stiffness has been linked to sport exposure and recent loading, which supports its use as a surveillance variable rather than a stand-alone diagnostic test [4].

Tips

- *Measure in the same time-window relative to training or treatment, because warm-up and early recovery effects can shift values within minutes to hours.*
- *Document recent exercise, stretching, massage, heat-cold exposure, and analgesic use before every session.*

Highlights

- *Neuromechanical biomarkers provide local, repeatable, non-invasive information about tissue mechanical state.*
- *Their value is greatest when they are interpreted together with symptoms, function, and exposure history rather than as isolated numbers.*

3.2 What the Variables Mean

3.2.1 Muscle Tone

Clinically, muscle tone is often described as resistance to passive stretch, but physiologically it emerges from interacting neural, reflex, contractile, and intrinsic viscoelastic mechanisms [5]. In the context of resting myotonometry, frequency is an operational index of intrinsic tension in the relaxed tissue rather than a complete measure of neural tone in the neurological sense [6]. Accordingly, a higher resting frequency should usually be interpreted as a more tensioned local tissue state, not as proof of increased motor-unit discharge by itself [7].

3.2.2 *Stiffness*

Stiffness is the resistance of a tissue to deformation under an applied force, but the reported quantity depends on the measurement model, so dynamic stiffness from myotonometry is not the same construct as shear modulus or converted elastic modulus from elastography [8]. At the muscle level, stiffness rises with passive stretch and active contraction because fascicle tension, intramuscular pressure, and connective-tissue loading all increase [9].

3.2.3 *Elasticity and Viscoelastic Behavior*

Elasticity refers to the ability of a tissue to recover shape after deformation, whereas viscoelastic behavior additionally includes time-dependent phenomena such as stress relaxation, creep, and hysteresis [8]. In Myoton-derived variables, decrement is inversely related to elasticity, while relaxation time and creep describe how rapidly and how progressively tissue recovers after the external impulse [10]. These variables matter because recovery interventions may alter not only absolute stiffness but also how quickly tissue dissipates or restores mechanical energy [11].

Practical Box

Myotonometric studies routinely separate frequency, stiffness, decrement, relaxation time, and creep because different interventions can move these variables in different directions [12]. Static stretching can reduce shear elastic modulus or hardness without implying that every component of tone or neuromuscular readiness has decreased proportionally [13].

Highlights

- *Tone is a broad physiological concept, but in device-based recovery work it must be operationalized carefully.*
- *Stiffness, elasticity, and viscoelasticity are related yet non-equivalent descriptors of tissue behavior.*

Tips

- *Define every reported variable in the methods section and state the unit explicitly, especially when mixing Hz, N/m, ms, Deborah number, m/s, kPa, and converted modulus.*
- *Avoid writing 'elasticity' when the instrument actually reports decrement or shear wave velocity, because these are not interchangeable physical quantities.*

Typical Values

- *In one reliability study, standard error of measurement for MyotonPRO frequency was about 0.3–0.8 Hz and for stiffness about 7.4–20.9 N/m across neck and orofacial muscles, which shows the scale of ordinary measurement noise [14].*
- *Minimum detectable change in the same study ranged from 0.8 to 2.2 Hz for frequency and from 20.5 to 57.9 N/m for stiffness, which is useful when setting decision thresholds [14].*

3.3 Biological Determinants of the Measured Signal

3.3.1 Neural and Cross-Bridge Contributions

Measured muscle mechanics are influenced by both active and passive elements, including residual cross-bridge attachment, calcium-dependent activation state, spindle-related reflex background, and the recent contractile history of the tissue [15]. Thixotropy is especially relevant in practice, because prior movement and recent contraction can change passive resistance even when electromyographic activity is minimal, which means the pre-measurement routine is part of the biology of the signal [2].

3.3.2 Titin and Sarcomeric Passive Tension

Titin is a dominant determinant of physiological passive stiffness in skeletal muscle, particularly as sarcomere length increases and passive tension develops [16]. This makes sarcomeric length control essential when comparing repeated measurements, because the same muscle can appear mechanically different simply because it is measured at another point on its length-tension curve [17].

3.3.3 Extracellular Matrix, Fascia, Fluid Shifts, and Temperature

The extracellular matrix, especially perimysial and endomysial collagen architecture and cross-linking, contributes substantially to passive biomechanical behavior and lateral force transmission [18]. Loading, aging, fibrosis, and chronic training can all remodel the matrix, so a mechanical biomarker can represent connective-tissue adaptation as much as myofibrillar state [19]. Acute fluid shifts also matter, because eccentric exercise can alter stiffness together with muscle thickness, and delayed-onset soreness studies show that edema-related changes do not behave identically across muscle, deep fascia, and different measurement modalities [20]. Tissue temperature is another confounder, because mechanical stiffness is temperature-sensitive and warm-up-related reductions in stiffness are one plausible mechanism linking preparation to improved movement performance [21].

Practical Box

Delayed onset muscle soreness (DOMS)-related experiments have shown that mechanical changes can occur together with increased muscle thickness or fascia stiffness, but the compartment showing the strongest relationship to pain may differ between studies [20]. Acute static stretching reduces passive fascicle stiffness and can shift slack angle, which demonstrates that short-term mechanical changes do not require structural remodeling [13].

Highlights

- *Measured stiffness is an emergent property of cross-bridge state, titin, extracellular matrix architecture, and fluid-temperature conditions.*
- *For this reason, a biomarker change is mechanistically interpretable only if posture, muscle length, and thermal state are known.*

Tips

- *Stabilize participants in a controlled ambient environment before testing, because temperature-sensitive tissues need time to equilibrate.*
- *Standardize pre-measurement rest and avoid repeated strong contractions immediately before baseline acquisition, because contractile history and thixotropy alter passive resistance.*

Typical Values

- *In healthy triceps surae, passive shear modulus rises as the ankle is dorsiflexed toward the slack-angle region and beyond, which confirms the strong length dependence of passive stiffness [9].*
- *Five minutes of static stretching was sufficient in one adolescent rectus femoris study to reduce stiffness while increasing blood flow, which shows that 'normal' values are highly time-dependent after warm-up maneuvers [22].*

3.4 How the Properties Are Measured

3.4.1 Hand-Held Myotonometry

Hand-held myotonometry applies a brief mechanical impulse to superficial tissue and derives frequency, dynamic stiffness, decrement, relaxation time, and creep from the resulting damped oscillations [8]. The method is portable, fast, and generally reliable for many superficial muscles and tendons, with recent syntheses reporting consistently high intra-rater and inter-rater reliability for frequency and stiffness across many sites [23]. Its main limitations are depth sensitivity, sensitivity to subcutaneous tissue thickness, and the fact that it samples a small local region rather than the whole muscle [24]. Recent evidence extends these findings to athletic populations, where a study [25] demonstrated high intra- and inter-rater reliability of MyotonPRO measurements across multiple sports disciplines, reinforcing its suitability for longitudinal monitoring in trained individuals.

3.4.2 Shear Wave Elastography

Shear wave elastography (SWE) estimates tissue mechanical behavior from the propagation speed of induced shear waves and offers better anatomical targeting than point myotonometry [1]. In skeletal muscle, shear wave velocity or shear modulus changes with both passive lengthening and contraction, which makes SWE useful for studying localized passive stiffness and task-dependent behavior [9]. Because skeletal muscle is anisotropic and violates simple isotropic assumptions, reporting shear wave velocity is often preferable to converting data into elastic modulus values that imply a more idealized material model than muscle actually satisfies [1, 21].

3.4.3 Convergent Validity and Non-equivalence

Myotonometry and SWE show moderate to strong correlations in several muscles, but they are not interchangeable because they interrogate different depths, different loading conditions, and different mathematical constructs [26, 27]. This non-equivalence is not a weakness if it is handled correctly, because the two methods can be used as complementary tools, with myotonometry optimized for rapid surveillance and SWE optimized for anatomically resolved interrogation [1].

Muscle mechanical behavior can also be examined with complementary methods that operate at different levels of biological organization [28]. Electromyography is useful for documenting muscle quiescence during passive protocols and for contextualizing activation-dependent contributions during active tasks, whereas dynamometry quantifies passive torque-angle behavior at the joint or muscle–tendon-unit level and therefore does not isolate the mechanical state of a single local tissue region [29, 30]. Magnetic resonance elastography provides a noninvasive MRI-based option for assessing skeletal muscle stiffness in deeper tissues and under loading conditions, while invasive isolated- or skinned-fiber preparations remain essential for resolving sarcomeric viscoelasticity and titin-related passive stiffness mechanisms [31]. These approaches should therefore be viewed as complementary rather than interchangeable with myotonometry or ultrasound elastography, because each samples a different construct, spatial scale, and source of variance [28].

Practical Box

MyotonPRO and SWE were moderately to strongly correlated in rectus femoris, biceps femoris, tibialis anterior, and medial gastrocnemius, but correlation strengths differed by muscle and by whether the tissue was at rest or contracting [26, 27]. Ultrasound elastography has also demonstrated validity against external mechanical standards and magnetic resonance elastography in muscle applications [32].

Highlights

- *Myotonometry is a rapid local surveillance tool for superficial tissues.*
- *SWE offers better spatial targeting, but its outputs depend strongly on acquisition assumptions and processing choices.*

Tips

- *For SWE, align the probe with fascicle direction and report whether you analyzed shear wave velocity, shear modulus, or converted elastic modulus.*
- *For MyotonPRO, keep the probe perpendicular to the skin, use the same landmark each session, and document subcutaneous thickness when possible.*

Typical Values

- *In healthy young adults, resting medial gastrocnemius values of about 278 N/m by MyotonPRO corresponded to about 12.3 kPa by SWE in one direct-comparison study, which illustrates that different devices can rank a muscle similarly while using different scales [27].*
- *During voluntary contraction in the same protocol, medial gastrocnemius stiffness increased to about 382 N/m by MyotonPRO and about 79.7 kPa by SWE, which confirms the marked activation sensitivity of both methods [27].*

3.5 Reliability, Validity, and Major Confounders

3.5.1 Reliability and Smallest Real Change

MyotonPRO reliability is usually strongest for resting measurements collected with fixed posture, fixed site, and the same operator, whereas reliability declines when joint position or voluntary contraction is not standardized [23, 33]. In a recent protocol paper, intraclass correlation coefficients for tissue stiffness, tone, and elasticity ranged from 0.52 to 0.97 across sites, which illustrates why smallest detectable change should be reported whenever repeated assessments are used for monitoring [34]. SWE can be reliable, but current literature still shows wide methodological heterogeneity and variable reliability values, particularly when probe alignment, region-of-interest selection, contraction level, or operator technique differ [1, 26].

3.5.2 The Dominant Confounders

The dominant confounders are muscle length or joint angle, contraction state, recent loading, temperature, body position, subcutaneous thickness, site selection, and probe orientation relative to fascicles [9, 17, 33]. Even within one muscle group, stiffness may vary regionally and across layers, so site marking and repeatable anatomical landmarks are mandatory if longitudinal change is the clinical target [11, 35]. As a rule, researchers should treat between-session changes smaller than the method's standard error of measurement or minimum detectable change as indeterminate rather than biological [14, 34].

Practical Box

A systematic review of forty-eight studies found consistently high MyotonPRO reliability across many muscles, but also emphasized methodological heterogeneity and gaps for some parameters and body regions [23]. A scoping review of skeletal muscle SWE similarly identified a lack of consensus and wide variability in protocols, particularly concerning age, sex, and methodological standardization [1].

Highlights

- *Reliability is a property of the protocol, not of the device name alone.*
- *Without SEM or MDC, longitudinal interpretation remains vulnerable to false positive findings [34].*

Tips

- *Fix joint angle, body position, contraction instructions, landmarking, operator, and number of repeated trials before comparing sessions or athletes.*

- *Record skinfold or subcutaneous thickness when using superficial devices over thicker soft-tissue regions, because apparent differences may reflect tissue depth rather than muscle biology [24].*

Typical Values

- *Across seven sites in a healthy protocol paper, ICCs for tissue stiffness, tone, and elasticity ranged from 0.52 to 0.97, which means some sites are monitoring-ready and others require more caution [34].*
- *In a knee-angle reliability study, MyotonPRO stiffness SEM ranged from 3.8 to 37.9 N/m between operators and from 7.9 to 52.0 N/m within operator across days, which gives a practical sense of expected variability [33].*

3.6 Athlete Applications and Interpretation

3.6.1 Acute Post-exercise Patterns

Exercise-induced changes are not uniform across protocols, because some eccentric-loading models show transient increases in stiffness, whereas others show decreases in dynamic stiffness or elastic modulus together with increased thickness, depending on muscle, timing, and device [20, 36]. This is why the athlete context matters, because a post-match or post-training value can reflect protective tone, transient edema, altered fascicle slackness, connective-tissue loading, or simple residual warm-up effects rather than a single injury mechanism [3].

3.6.2 Chronic Training Adaptation

Chronic training can also reshape mechanical profiles, with evidence that amateur basketball players have greater gastrocnemius-Achilles complex stiffness than non-athletes and that professional soccer players can show higher rectus femoris stiffness but lower patellar and quadriceps tendon stiffness than sedentary controls [4]. Sport-specific range-of-motion exposure also appears relevant, because elite athletes in flexibility- and range-of-motion-demanding sports can show altered hamstring shear modulus profiles compared with controls [37].

3.6.3 Practical Interpretation Rules

For applied monitoring, the most defensible strategy is to build athlete-specific baselines under tightly standardized conditions and then interpret changes alongside pain, range of motion, performance, and exposure history rather than as isolated numbers [34]. Population reference values are still useful for orientation, but within-athlete trajectories are usually more actionable because absolute values remain strongly muscle-specific and protocol-specific [1]. An increase in stiffness with preserved function may represent adaptive readiness in one context, whereas the same change with pain, swelling, or range-of-motion loss may indicate incomplete recovery in another context [4].

Practical Box

Amateur basketball players showed greater gastrocnemius-Achilles complex stiffness than non-athletes, which is consistent with habitual loading adaptation [4]. Professional soccer players showed higher rectus femoris stiffness but lower patellar and quadriceps tendon stiffness than sedentary controls, which demonstrates that adaptation is tissue-specific rather than globally 'stiffer' or 'softer' [38].

Highlights

- *Post-exercise stiffness responses are pattern-based, not universal.*
- *Chronic training adaptations are sport-specific and tissue-specific.*

Tips

- *Build an athlete's baseline from repeated low-noise sessions and compare future data against that personal reference, not against a single published mean.*
- *Interpret abnormal changes together with soreness, range of motion, performance, and recent exposure to separate adaptive readiness from incomplete recovery.*

Typical Values

- *Healthy controls in one gastrocnemius study showed medial gastrocnemius stiffness around 281 plus or minus 29 N/m by MyotonPRO, which sits very close to values reported in other young-adult protocols and therefore provides a reasonable orientation point for lower-leg monitoring [24].*
- *In one standardized protocol, resting tibialis anterior stiffness was substantially higher than gastrocnemius or thigh-muscle stiffness, which again shows that local reference ranges must remain muscle-specific [27].*

3.7 Minimal Reporting Set for Research and Practice

A publishable or clinically interpretable neuromechanical dataset should report the tissue and exact site, side, body position, joint angle, muscle state, time since exercise, ambient thermal conditions, device settings, operator, number of trials, and reliability statistics or minimum detectable change [34]. Without this metadata, cross-study synthesis becomes weak and apparent intervention effects can reflect protocol drift more than tissue change [23].

Practical Box

Protocol papers that combine landmark charts, repeated-rater designs, and minimum detectable change tables are more clinically useful than papers that only report pre-post means [23]. Shear wave elastography reviews repeatedly identify missing acquisition details as a major source of inconsistency across studies [1].

Highlights

- *Measurement standardization is part of the intervention science, not an administrative afterthought.*
- *The best endpoint is the one whose biological target and technical error are both known.*

Tips

- *Use a fixed collection checklist so that site, angle, warm-up status, and operator do not drift over a season or a trial.*
- *Where possible, pair a fast surveillance tool with one anatomically resolved method for calibration in a subsample.*

Typical Values

- *For many superficial muscles, a change much smaller than about 20–60 N/m on MyotonPRO may fall inside ordinary measurement error depending on site and protocol, so decisions should not be based on trivial single-session shifts [1].*
- *Likewise, wide published ranges of resting shear wave velocity or kilopascal values across muscles and protocols mean that standardized within-laboratory reference data are often more informative than literature-wide averages [1].*

References

1. Stiver ML, Mirjalili SA, Agur AMR (2023) Measuring shear wave velocity in adult skeletal muscle with ultrasound 2-D shear wave elastography: a scoping review. Ultrasound Med Biol 49:1353–1362
2. Simons GD, Mense S (1998) Understanding and measurement of muscle tone as related to clinical muscle pain. Pain 75:1–17
3. Ličen U, Kozinc Ž (2022) Using shear-wave elastography to assess exercise-induced muscle damage: a review. Sensors 22:7574
4. Chang T-T, Li Z, Wang X-Q, Zhang Z-J (2020) Stiffness of the gastrocnemius-Achilles tendon complex between amateur basketball players and the non-athletic general population. Front Physiol 11
5. Ganguly J, Kulshreshtha D, Almotiri M, Jog M (2021) Muscle tone physiology and abnormalities. Toxins (Basel) 13:282
6. Masi AT, Hannon JC (2008) Human resting muscle tone (HRMT): narrative introduction and modern concepts. J Bodyw Mov Ther 12:320–332
7. Nguyen AP, Detrembleur C, Fisette P, Selves C, Mahaudens P (2022) MyotonPro is a valid device for assessing wrist biomechanical stiffness in healthy young adults. Front Sports Act Living 4
8. McGowen JM et al (2023) The utility of myotonometry in musculoskeletal rehabilitation and human performance programming. J Athl Train 58:305–318
9. Chernak LA, DeWall RJ, Lee KS, Thelen DG (2013) Length and activation dependent variations in muscle shear wave speed. Physiol Meas 34:713–721

10. Bohlen L et al (2022) Effect of osteopathic techniques on human resting muscle tone in healthy subjects using myotonometry: a factorial randomized trial. Sci Rep 12:16953
11. Kawczyński A et al (2018) Trapezius viscoelastic properties are heterogeneously affected by eccentric exercise. J Sci Med Sport 21:864–869
12. Schneider S, Peipsi A, Stokes M, Knicker A, Abeln V (2015) Feasibility of monitoring muscle health in microgravity environments using Myoton technology. Med Biol Eng Comput 53:57–66
13. Hirata K, Kanehisa H, Miyamoto N (2017) Acute effect of static stretching on passive stiffness of the human gastrocnemius fascicle measured by ultrasound shear wave elastography. Eur J Appl Physiol 117:493–499
14. Taş S, Yaşar Ü, Kaynak BA (2021) Interrater and intrarater reliability of a handheld myotonometer in measuring mechanical properties of the neck and orofacial muscles. J Manip Physiol Ther 44:42–48
15. Shenkman BS, Tsaturyan AK, Vihlyantsev IM, Kozlovskaya IB, Grigoriev AI (2021) Molecular mechanisms of muscle tone impairment under conditions of real and simulated space flight. Acta Nat 13:85–97
16. Brynnel A et al (2018) Downsizing the molecular spring of the giant protein titin reveals that skeletal muscle titin determines passive stiffness and drives longitudinal hypertrophy. Elife 7
17. Hirata K, Kanehisa H, Miyamoto-Mikami E, Miyamoto N (2015) Evidence for intermuscle difference in slack angle in human triceps surae. J Biomech 48:1210–1213
18. Lieber RL, Meyer G (2023) Structure-function relationships in the skeletal muscle extracellular matrix. J Biomech 152:111593
19. Wohlgemuth RP, Brashear SE, Smith LR (2023) Alignment, cross linking, and beyond: a collagen architect's guide to the skeletal muscle extracellular matrix. Am J Phys Cell Phys 325:C1017–C1030
20. Kisilewicz A et al (2020) Eccentric exercise reduces upper trapezius muscle stiffness assessed by shear wave elastography and myotonometry. Front Bioeng Biotechnol 8
21. Kruse SA et al (2000) Tissue characterization using magnetic resonance elastography: preliminary results. Phys Med Biol 45:1579–1590
22. Caliskan E et al (2019) Effects of static stretching duration on muscle stiffness and blood flow in the rectus femoris in adolescents. Med Ultrason 21:136
23. Lettner J et al (2024) Evaluating the reliability of MyotonPro in assessing muscle properties: a systematic review of diagnostic test accuracy. Medicina (B Aires) 60:851
24. Fröhlich-Zwahlen AK, Casartelli NC, Item-Glatthorn JF, Maffiuletti NA (2014) Validity of resting myotonometric assessment of lower extremity muscles in chronic stroke patients with limited hypertonia: a preliminary study. J Electromyogr Kinesiol 24:762–769
25. Trybulski R et al (2024) Reliability of MyotonPro in measuring the biomechanical properties of the quadriceps femoris muscle in people with different levels and types of motor preparation. Front Sports Act Living 6
26. Kelly JP et al (2018) Characterization of tissue stiffness of the infraspinatus, erector spinae, and gastrocnemius muscle using ultrasound shear wave elastography and superficial mechanical deformation. J Electromyogr Kinesiol 38:73–80
27. Lee Y, Kim M, Lee H (2021) The measurement of stiffness for major muscles with shear wave elastography and Myoton: a quantitative analysis study. Diagnostics 11:524
28. Bilston LE, Tan K (2015) Measurement of passive skeletal muscle mechanical properties in vivo: recent progress, clinical applications, and remaining challenges. Ann Biomed Eng 43:261–273
29. Magnusson SP (1998) Passive properties of human skeletal muscle during stretch maneuvers. Scand J Med Sci Sports 8:65–77
30. Raiteri BJ, Hug F, Cresswell AG, Lichtwark GA (2016) Quantification of muscle co-contraction using supersonic shear wave imaging. J Biomech 49:493–495
31. Dresner MA et al (2001) Magnetic resonance elastography of skeletal muscle. J Magn Reson Imaging 13:269–276

32. Chino K, Akagi R, Dohi M, Fukashiro S, Takahashi H (2012) Reliability and validity of quantifying absolute muscle hardness using ultrasound elastography. PLoS One 7:e45764
33. Chen G et al (2019) Reliability of a portable device for quantifying tone and stiffness of quadriceps femoris and patellar tendon at different knee flexion angles. PLoS One 14:e0220521
34. Muckelt PE et al (2022) Protocol and reference values for minimal detectable change of MyotonPRO and ultrasound imaging measurements of muscle and subcutaneous tissue. Sci Rep 12:13654
35. Belghith K, Zidi M, Fedele J-M, Bou Serhal R, Maktouf W (2023) Spatial distribution of stiffness between and within muscles in paretic and healthy individuals during prone and standing positions. J Biomech 161:111838
36. Agten CA et al (2017) Delayed-onset muscle soreness: temporal assessment with quantitative MRI and shear-wave ultrasound elastography. Am J Roentgenol 208:402–412
37. Avrillon S et al (2020) Hamstring muscle elasticity differs in specialized high-performance athletes. Scand J Med Sci Sports 30:83–91
38. Taş S et al (2020) Knee muscle and tendon stiffness in professional soccer players: a shear-wave elastography study. J Sports Med Phys Fitness 60

Chapter 4
Thermal Interventions: Cold and Heat

Abstract This chapter examines thermal interventions as dose-controlled recovery tools, with emphasis on how cold, heat, and contrast exposures influence vascular function, microcirculation, neural signaling, pain modulation, and recovery-related physiological processes. It guides the reader from foundational mechanisms to practical implementation, explaining how thermal dose is defined, delivered, and measured, and why the biological response may differ from the applied stimulus. The chapter also details the main neurophysiological and microvascular pathways underlying cold- and heat-induced effects, highlights dose–response logic across athletic and clinical contexts, and discusses when these strategies may support recovery or interfere with adaptation. Particular attention is given to safety, measurement validity, methodological rigor, and interpretation of common monitoring variables. Throughout, the chapter integrates mechanistic evidence with applied examples, key take-home points, technical guidance, and typical values to support precise, evidence-based use of thermal interventions in practice.

4.1 Thermal Dose: What It Is, How to Measure It, and Why It Matters

Thermal interventions are best treated as dose-controlled biological perturbations that alter perfusion, neural excitability, and cellular stress signaling, thereby changing pain and performance recovery trajectories in ways that can be measured [1]. The goal is not cold versus heat, but matching thermal dose to the desired microvascular, neuromechanical, and nociceptive endpoints while respecting contraindications and risk [1]. Because most athletic "recovery" outcomes are time-dependent (hours–days), thermal protocols should be described using a dose–response logic rather than a single recipe [2].

R. Trybulski, *Biomarker-Guided Physical Recovery Interventions*,
SpringerBriefs in Forensic and Medical Bioinformatics,
https://doi.org/10.1007/978-981-92-0631-5_4

Practical Box

Systematic reviews show cold-water immersion is widely used post-exercise and can improve soreness and some performance-related recovery outcomes in physically active populations [3]. Evidence also indicates that frequent post-resistance-training cold exposure can attenuate hypertrophy-related signaling and longer-term benefits, so "recovery" benefits must be weighed against adaptation goals [4].

Highlights

- *Treat temperature as a dose (magnitude × duration × surface area × medium) and link it to a biological endpoint.*
- *Use thermal modalities strategically because "better short-term recovery" can conflict with "better long-term adaptation" depending on training phase.*

A practical definition of thermal dose is the externally applied temperature gradient (relative to baseline), multiplied by exposure duration and exposed surface area, with strong dependence on the delivery medium (water, air, contact) [2]. Water immersion also adds hydrostatic pressure and often changes convective heat transfer compared with air, so thermal interventions may carry mechanical confounding that must be acknowledged [5]. The tissue effect (skin, muscle, core temperature change) can differ substantially between individuals because perfusion and insulation modify heat exchange, making direct temperature measurement preferable to assuming delivered dose equals received dose [6].

For example, recent randomized evidence indicates that the duration component of cold dose is not linearly beneficial, because ankle-applied cold compression produced progressively greater vasoconstriction with longer exposure, whereas hypoalgesic effects plateaued by approximately 10 min and cold-induced vasodilation plus post-cooling hyperaemia became more pronounced at 15–20 min, supporting the existence of a pragmatic therapeutic window rather than a simple more-time-more-effect model [7].

Consistent with this threshold logic, forearm-based contrast heat–cold compression research in mixed martial arts athletes showed that 10 min was sufficient to induce significant changes in muscle biomechanics, pain threshold, perfusion, and force, even though some variables improved further at 20 min, suggesting that the minimal effective duration may be shorter than traditional protocols assume for several recovery targets [8].

Practical Box

Network meta-analytic comparisons suggest different cold-water immersion temperature–duration combinations may differentially relate to outcomes like soreness versus creatine kinase and neuromuscular recovery [9]. Applied studies in combat sport contexts have used contrast heat–cold–pressure devices and tracked concurrent changes in perfusion units and neuromechanical variables, illustrating the need to record all stimulus components [10].

Highlights

- *"Thermal dose" is not a brand name or modality label; it is a quantifiable stimulus with multiple parameters.*
- *When the medium is water, do not ignore hydrostatic pressure as a plausible co-stimulus.*

Tip

- *Record baseline skin temperature and at least one post-exposure skin temperature point to confirm that the intended cooling/heating magnitude occurred.*
- *If comparing modalities, report both exposure conditions and the achieved skin temperature because efficacy can depend on reaching a cooling threshold.*

Typical Values

- *In the recovery literature, cold-water immersion is commonly defined as water temperatures < 15 °C, but actual protocols vary [11].*
- *Meta-analytic dose work often finds 10–15 min immersion durations are frequently associated with favorable recovery effects, though context matters [2].*
- *Whole-body cryostimulation studies sometimes operationalize an "analgesic threshold" as achieving skin temperatures below ~ 13.6 °C, emphasizing response-based dosing [6].*

4.1.1 Measuring Temperature Exposure and Response

Core temperature is best approximated by criterion methods such as rectal temperature, while some alternative devices can show systematic bias during exercise hyperthermia [12]. Ingestible telemetric pills provide a practical, validated approach for estimating gastrointestinal temperature as a surrogate of core temperature in exercise contexts when used with appropriate methodological controls [13]. Multiple validation studies support acceptable agreement between ingestible systems and rectal reference measures under controlled conditions, while emphasizing device- and context-specific error [14].

Timing of pill ingestion (within common practical windows) does not appear to meaningfully alter validity as an index of core temperature during exercise-heat stress in available evidence [15]. Aural/tympanic measures can substantially underestimate core temperature in hyperthermic exercising individuals, which is critical when safety decisions depend on temperature thresholds [16]. Clinical evidence specifically cautions against aural thermometry as a diagnostic substitute for rectal thermometry in exertional heat stroke contexts [17].

Skin temperature can be measured using contact thermistors or infrared systems, but validity and reliability depend on device type, anatomical site, and environmental conditions [18]. Infrared thermography reliability varies by body region and methodology, so standardized acquisition procedures are necessary for meaningful longitudinal interpretation [19].

Practical Box

Athletic studies increasingly explore infrared thermography responses to exercise as a non-invasive proxy for local thermal/perfusion changes, but relationships with internal load biomarkers remain heterogeneous [20]. Specific thermography procedures in athletes can show excellent intra- and inter-session reliability when acquisition is standardized, illustrating that method quality is decisive [21].

Highlights

- *Use core-temperature methods that match the risk profile, because some peripheral devices systematically underestimate hyperthermia.*
- *Skin temperature is useful for confirming local cooling/heating, but it is not equivalent to muscle or core temperature.*

Tip

- *For ingestible pills, follow published methodological guidance on ingestion, calibration checks, and logging to avoid artifact from recent fluid ingestion and transit issues [13].*
- *For infrared thermography, standardize room conditions, acclimation time, camera distance/angle, emissivity assumptions, and anatomical ROI selection to reduce noise [19].*
- *For heat-illness risk decisions, avoid substituting aural methods for rectal methods because misclassification can be clinically catastrophic [17].*

Typical Values

- *In validation work, ingestible pill temperature is often treated as a close surrogate to rectal temperature under controlled protocols, but any single device can show context-dependent bias [14].*
- *Aural temperature tends to underestimate rectal core temperature increasingly as hyperthermia becomes more severe [16].*

4.2 Cold Interventions (Cryotherapy Continuum): Vascular Mechanisms, Neurophysiology, and Dose–Response

4.2.1 Vascular and Microvascular Mechanisms of Cold Exposure Relevant to Recovery

Cold exposure rapidly increases sympathetic vasoconstrictor drive in cutaneous beds, reducing peripheral blood flow and skin temperature to limit heat loss [22]. Reduced cutaneous perfusion and local tissue cooling can decrease transcapillary fluid flux and may influence post-exercise edema dynamics, though effects depend strongly on timing, exercise type, and tissue depth cooled [11].

In distal extremities, prolonged cold exposure can produce cyclic “cold-induced vasodilation” episodes superimposed on vasoconstriction, reflecting complex local and neural control rather than a simple monotonic constriction model [22]. Cold-induced vasodilation responses show within-subject variability even under standardized laboratory conditions, which matters if using cold-provocation vascular tests as biomarkers [23].

Practical Box

Contrast-enhanced ultrasound work has been used to quantify changes in intramuscular microvascular blood flow after cycling and systemic cold-water exposure, illustrating that microvascular endpoints are measurable but modality- and tissue-specific [24]. Applied athlete recovery settings increasingly combine cooling with compression, which can amplify perfusion changes and complicate attribution to temperature alone [8].

Highlights

- *Cold dose modifies perfusion primarily through sympathetic vasoconstriction, but distal beds may show paradoxical cyclic vasodilation.*
- *Microvascular response biomarkers can be variable, so repeated-measures designs and standardized protocols are essential for inference.*

Tip

- *If perfusion is an endpoint, control posture, ambient temperature, and acclimation time because these factors modulate cutaneous vasomotor tone and can obscure intervention effects.*
- *When using immersion, specify immersion depth and limb position because local hydrostatic pressure and convective cooling depend on geometry [5].*

Typical Values

- *Cold-induced vasodilation research frequently uses prolonged digit immersion at very cold temperatures (e.g., 4 °C) to provoke oscillatory responses, but reproducibility is imperfect [23].*
- *Recovery-oriented cold water immersion protocols more commonly use less extreme whole-body or limb water temperatures than classic cold-induced vasodilation tests, reinforcing that "cold" is not a single physiological condition [2].*

4.2.2 *Neurophysiological Mechanisms: TRP Channels, Nerve Conduction, and Pain Modulation*

Cold sensation and aspects of cold-induced analgesia are mediated in part by temperature-sensitive transient receptor potential (TRP) channels, including TRPM8, which acts as a principal transducer of innocuous cooling in sensory neurons [25]. Experimental pain work shows that TRPM8 activation can produce clinically relevant analgesia in chronic pain states, supporting a plausible biological pathway for cold-related pain modulation [26].

Cryotherapy can reduce peripheral nerve conduction velocity as local skin temperature decreases, providing a mechanistic basis for increased pain threshold and tolerance after cooling [27]. Because nerve conduction and receptor kinetics are temperature-dependent, the magnitude of analgesia depends on achieving sufficient tissue cooling rather than merely feeling cold [6].

Practical Box

Whole-body cryostimulation dosing studies explicitly target a skin-temperature threshold to optimize analgesic effects, illustrating response-based dosing rather than fixed time-only prescriptions [6]. Combat-sport crossover work has reported acute changes in neuromechanical and perceptual measures after brief ice-massage interventions during rest breaks, supporting feasibility of short, targeted cooling between bouts [28].

Highlights

- *Cold can modulate pain via TRP-mediated sensory transduction and via reduced nerve conduction velocity, which are dose-dependent.*
- *Analgesia is a physiological outcome that requires achieving sufficient cooling, so measure skin temperature when possible.*

Tip

- *If the intended mechanism is analgesia, measure pressure pain threshold (or another validated nociceptive endpoint) and relate it to achieved skin temperature.*

- *Avoid assuming equivalence across modalities (ice pack, cold water immersion, whole-body cryotherapy) unless you can show similar tissue cooling, because nerve conduction changes scale with temperature reduction.*

Typical Values

- *A proposed target for cold-induced analgesia is achieving skin temperatures below 13.6 °C in some cryostimulation contexts [6].*
- *Cooling-related nerve conduction slowing has been demonstrated alongside increased pain threshold and tolerance in controlled studies [27].*

4.2.3 Dose–Response Evidence in Athletes: What Cold Improves, and When It Can Be Counterproductive

Systematic reviews and meta-analyses generally support cold-water immersion as effective for reducing perceived muscle soreness after strenuous exercise compared with passive recovery, though effect sizes and certainty vary [11]. Meta-analytic dose analyses suggest immersion time (often 10–15 min) and temperature interact with outcomes, reinforcing the need to prescribe rather than genericize ice baths [2].

In team-sport contexts, systematic review evidence supports that cold water immersion and contrast water therapy can improve certain recovery outcomes, but the magnitude depends on outcome selection and study quality [5]. At the same time, mechanistic and longitudinal evidence indicates routine post-resistance-training cold water immersion can attenuate anabolic signaling and may reduce long-term hypertrophy and strength gains [4]. Therefore, prioritize it in congestion phases where short-term availability is the goal, and reduce routine use during hypertrophy-focused mesocycles [1].

Practical Box

A meta-analysis report cold water immersion is often superior to other recovery methods for soreness, while being similar for some performance outcomes, underscoring outcome-specific benefits [29]. Whole-body cryotherapy has been widely adopted, yet Cochrane-level evidence notes insufficient evidence for benefit and limited adverse-event surveillance, making safety documentation essential [30].

Highlights

- *Cold is most consistently supported for soreness-related outcomes, with more variable evidence for objective performance measures.*
- *Frequent post-lifting cold exposure can blunt hypertrophy adaptation signals, so more recovery is not always more progress [4].*

Tip

- *Stratify prescriptions by training goal (match congestion vs. hypertrophy block) and document rationale to avoid "one-size-fits-all" recovery.*
- *Report or monitor adverse events and tolerability, because many trials do not actively surveil harms despite plausible risks.*

Typical Values

- *Cold-water immersion in DOMS/recovery research is commonly < 15 °C, with many protocols clustering around moderate durations [11].*
- *Dose-focused syntheses frequently highlight 10–15 min as a common and often effective duration range, though the optimal temperature may differ by outcome [2].*

Practical Guideline for Cold-Water Immersion (CWI)

- *Use CWI when the priority is short-turnaround recovery, especially after matches, repeated high-intensity efforts, travel, or sessions with substantial soreness and fatigue, because CWI is consistently supported for reducing muscle soreness and is often as good as or better than many other common recovery methods [29].*
- *A practical default prescription is 10–15 °C for 10–15 min, with the exercised muscle groups fully immersed, because this temperature–time range has the strongest dose-response support in the recovery literature [2].*
- *Use a thermometer, set the water before the athlete enters, start timing only once the target depth is reached, and keep the athlete seated or supported so the exposure is controlled and reproducible.*

- *Use CWI immediately after exercise or within the first recovery window when the goal is short-term availability for the next session or competition.*
- *Do not allow breath-holding or face immersion, because cold water can provoke strong autonomic responses and, in some situations, arrhythmogenic autonomic conflict.*
- *Stop the exposure if the athlete reports chest pain, severe dizziness, panic, unusual shortness of breath, or escalating distress, because CWI is a real cardiovascular stressor rather than a neutral comfort intervention.*

Practical Guideline for Cryotherapy (Whole-Body WBC or Local Cold Application)

- *Use cryotherapy when the main target is analgesia, perceptual recovery, or short-term soreness reduction, especially when a supervised cryotherapy facility is available, but remember that the evidence base is less standardized than for CWI [*31*].*
- *A practical whole-body cryotherapy exposure is usually 1–3 min in very cold, dry air, with protocols across studies typically ranging from about − 110 to − 140 °C, although the broader literature spans more extreme chamber settings [*32*].*
- *The athlete should enter fully dry, wearing the facility's required protective items such as gloves, socks/shoes, ear protection, and mouth/nose protection, because exposed wet skin and inadequate protection increase cold-injury risk.*
- *Use WBC only in a screened and supervised setting with clear stop rules, because published adverse events are uncommon but are considered largely preventable when contraindications and safety procedures are followed.*
- *If the goal is mainly short-term analgesia, local cooling around 10 min is often enough, because analgesic effects can occur without needing prolonged exposure [*33*].*
- *Do not send an athlete directly into explosive, accuracy-dependent, or high-skill activity immediately after substantial cooling, because cryotherapy can acutely reduce strength, dexterity, sprint ability, and agility in some contexts.*
- *If cryotherapy is used close to play, keep the exposure brief and insert a progressive re-warm-up before return to activity.*

4.3 Heat Interventions (Thermotherapy): Vascular Mechanisms, Cellular Stress Signaling, and Applied Dosing

4.3.1 Vascular and Endothelial Mechanisms of Heating Relevant to Recovery and Remodeling

Heat exposure increases skin blood flow and can lower peripheral vascular resistance, altering shear stress stimuli that are central to endothelial function [34]. Repeated passive heat therapy can improve endothelial function and reduce arterial stiffness and blood pressure in sedentary humans, supporting a vascular-adaptation pathway that overlaps with exercise mechanisms [35]. Mechanistic reviews emphasize nitric oxide bioavailability, shear-stress signaling, and heat-induced vascular remodeling as candidate mediators of heat therapy's cardiovascular effects [36].

Practical Box

Hot-water immersion training-style protocols have been studied as a feasible passive heating approach that can drive cardiovascular strain and adaptation-like responses when repeated [36]. Systematic reviews of passive heat therapies summarize multi-system benefits but also highlight that protocol heterogeneity makes best dose identification difficult without careful reporting [37].

Highlights

- *Heating is not only relaxing; it is a vascular stimulus that can drive endothelial adaptation signals with repeated exposure.*
- *Heat can be a deliberate tool for manipulating perfusion and shear-stress biology.*
- *Sauna research commonly emphasizes individualized tolerable exposures and warns against extremes, making safe dose contingent on person and context [38].*
- *Heat therapy trials often apply repeated sessions across weeks, indicating that chronic dosing (not single exposure) is central to vascular adaptation claims [35].*

Tip

- *Monitor blood pressure and symptoms in passive heating sessions, because vasodilation and cardiovascular load can be clinically meaningful even without exercise.*
- *If using hot-water immersion or sauna-like exposures, document ambient/medium temperature, duration, and cooling intervals for reproducibility and safety.*

4.3.2 Neurophysiological and Tissue-Mechanical Effects of Heat Relevant to Pain and Function

Superficial heat therapy can reduce pain and improve flexibility and function in musculoskeletal conditions, supporting a pragmatic role for heat in symptom modulation [39]. Thermotherapy's analgesic effects are plausibly mediated by increased local perfusion, reduced muscle spasm, and altered nociceptor sensitivity, though mechanisms likely vary by tissue and condition [39]. Because TRPV1 is a major molecular integrator for noxious heat and chemical stimuli, heat can influence sensory signaling pathways that overlap with pain modulation biology [40].

Practical Box

Applied contrast heat–cold–pressure protocols in athletes have been associated with acute changes in pain threshold and muscle biomechanical properties, illustrating combined thermal–mechanical effects [10]. Reviews of superficial heat emphasize short-term pain relief and functional improvements, which can be operationally valuable in return-to-training contexts [39].

Highlights

- *Heat can be positioned as a tool for pain and stiffness modulation, not just a comfort intervention.*
- *Thermosensory channels such as TRPV1 provide a plausible molecular bridge between thermal exposure and nociceptive signaling.*
- *Evidence summaries support that continuous low-level heat can provide pain relief and functional benefit, implying that mild but sustained heating is a meaningful dosing paradigm [39].*

- *Heat-sensitive TRP biology is triggered across specific thermal ranges, reinforcing that temperature magnitude matters rather than the label heat therapy [41].*

Tip

- *Use standardized pain endpoints (e.g., pressure pain threshold) and report measurement conditions, because pain is sensitive to context and expectancy.*
- *For superficial heat, document device temperature and application duration because thermal dose at the skin is what drives local effects.*

Practical Guidelines for Heat Therapy

- *Use heat when the target is stiffness reduction, pain relief, re-warming, or passive heat stress/adaptation, but do not treat it as an automatic post-exercise recovery tool because acute performance-recovery evidence remains mixed [42].*
- *A practical starting protocol is hot-water immersion at 40 °C for 20 min after the final session of the day, because this format has been implemented successfully in elite athletes in applied contexts [43].*
- *Where monitoring is available and a stronger heat dose is desired, 41 °C hot-water immersion can be used, and one experimental model found benefits when core temperature was maintained for about 25 min between 38.5 and 39.0 °C [44].*
- *A practical sauna option is 20 min of infrared sauna around the low-40 °C range, which improved next-day neuromuscular recovery and soreness in basketball players [45].*
- *Do not use heat if the athlete is still clearly hyperthermic, dizzy, or suspected of heat illness, because that scenario requires cooling and medical assessment, not additional heating.*
- *Heat therapy has recognized cardiovascular contraindications, including unstable cardiovascular states such as unstable angina, recent myocardial infarction, and severe aortic stenosis [38].*

4.4 Contrast (Hot–Cold) Protocols: Oscillatory Vascular Logic and What the Evidence Actually Supports

Contrast protocols are typically justified by alternating vasoconstriction and vasodilation to create oscillatory changes in blood flow that may influence metabolite clearance and edema dynamics [46]. However, mechanistic certainty is limited because many contrast studies have high risk of bias and heterogeneous protocols, complicating inference about the causal "pumping" mechanism [46]. Because contrast protocols in water also add immersion-related hydrostatic pressure, they can represent combined thermal–mechanical dosing even when framed as purely thermal [5].

Practical Box

Team-sport meta-analytic evidence has evaluated both cold water immersion and contrast water therapy, supporting that both are used in elite settings but with variable outcome effects [5]. Athlete trials using device-based contrast heat–cold with pressure have quantified perfusion and neuromechanical endpoints, illustrating biomarker-aligned contrast logic [10].

Highlights

- *Contrast protocols have a plausible vascular logic, but evidence quality and protocol variability remain key limitations.*
- *In immersion contexts, contrast is rarely thermal-only, so report pressure and immersion depth.*
- *Contrast water therapy trials vary widely, and systematic reviews emphasize that protocol heterogeneity limits firm conclusions about optimal cycle structure [46].*
- *In applied sport scenarios, contrast interventions are commonly used alongside other recovery modalities, reinforcing the need to isolate or explicitly describe multimodal stacks [5].*

Tip

- *Specify hot and cold phase temperatures, durations, number of cycles, and transition time, because these define the oscillatory dose.*

- *If using a combined heat–cold–pressure device, treat pressure as a co-dose and measure perfusion endpoints if mechanistic claims rely on microcirculation.*

Practical Guidelines for Contrast Therapy

- *Use contrast therapy when the athlete prefers alternating thermal exposure or when you want a mixed perceptual recovery stimulus, but note that the evidence suggests it is better than passive rest yet not consistently superior to CWI [46].*
- *A practical and well-studied contrast protocol is equal-time hot and cold cycles of about 1 min each, with a total duration of 10–15 min [47].*
- *A useful field version is hot water at 38 °C and cold water at 15 °C, alternating every minute [47].*
- *Across the wider literature, hot baths have ranged roughly from 35.5 to 45 °C and cold baths from 8 to 15 °C, but narrower mid-range targets are easier to standardize and usually more tolerable [46].*
- *Do not use contrast in an athlete who is presyncopal, heat-ill, cold-intolerant, or unable to tolerate either end of the thermal range safely.*
- *If the goal is the most evidence-supported cold recovery strategy, choose CWI rather than contrast [2, 5].*

4.5 Safety and Contraindications: Cold, Heat, and the Non-negotiables

4.5.1 Cold Exposure Risks: Autonomic Conflict, Arrhythmias, and Cardiovascular Strain

Cold water immersion can trigger powerful autonomic reflexes, including cold-shock responses and diving-related bradycardic reflexes, which can conflict and provoke arrhythmias in susceptible conditions [48]. Acute and prolonged cold exposure increases cardiovascular strain, and underlying cardiovascular disease can modify risk and hemodynamic responses [49]. Therefore, cold protocols should avoid breath-holding and face immersion in non-controlled contexts, because these amplify parasympathetic diving responses and increase autonomic conflict potential [48].

Practical Box

Reviews of voluntary cold-water exposure summarize both proposed benefits and risks, emphasizing that safety screening and gradual exposure matter [50]. Controlled observations show cold immersion can induce hemodynamic changes, reinforcing that recovery baths are cardiovascular stressors, not neutral events [51].

Highlights

- *Cold exposure can be arrhythmogenic in some contexts due to competing autonomic reflexes.*
- *Cold increases cardiovascular load, so screening and conservative dosing are essential in mixed clinical–sport environments.*
- *"Cold shock" and "diving" reflex interactions are most relevant when cold water and breath/face stimuli coexist, which is avoidable by protocol design.*
- *People with cardiovascular disease can show altered responses to cold exposure, supporting conservative dosing and medical oversight when indicated [49].*

Tip

- *Screen for cardiovascular disease risk and avoid protocols that combine cold with breath-holding/face immersion outside controlled supervision.*
- *Track symptoms (dizziness, chest discomfort, syncope) and measure blood pressure when appropriate because cold can elevate cardiovascular strain.*

4.5.2 Heat Exposure Risks: Syncope, Cardiovascular Contraindications, and Heat-Illness Measurement Errors

Sauna bathing and other heat exposures produce acute cardiovascular and hormonal changes that are generally tolerated by healthy individuals but have specific contraindications in unstable cardiovascular disease states [38]. Reviews identify contraindications such as unstable angina, recent myocardial infarction, and severe aortic stenosis, supporting a conservative safety screen before high-heat protocols

[38]. In exertional heat illness contexts, inaccurate temperature measurement can be life-threatening, and aural thermometry should not be used as a substitute for rectal thermometry for diagnosing exertional heat stroke [17].

Practical Box

Systematic reviews of sauna and heat exposure highlight potential benefits but emphasize incomplete adverse-event reporting and the need for better dose definition [37]. Meta-analysis demonstrates that aural temperature underestimates rectal core temperature in hyperthermic exercising individuals, supporting strict measurement standards when safety is on the line [16].

Highlights

- *Heat protocols require screening for cardiovascular contraindications and clear stop-rules for symptoms.*
- *If the use-case includes heat illness risk, measurement validity (rectal vs. peripheral) is a safety-critical issue, not a methodological detail.*
- *Sauna safety guidance emphasizes avoiding extremes and tailoring exposure duration and temperature to the individual, especially in clinical populations.*
- *Peripheral thermometry underestimation of core temperature becomes more problematic as hyperthermia increases, reinforcing rectal measurement for high-stakes decisions [16].*

Tip

- *Use validated core temperature methods when making clinical safety decisions, and do not apply correction factors to make aural readings approximate rectal values in exertional heat stroke scenarios.*
- *Document the heating protocol precisely (temperature, duration, cooling intervals) and ensure hydration and supervised exits to reduce syncope risk in vulnerable individuals.*

References

1. Petersen AC, Fyfe JJ (2021) Post-exercise cold water immersion effects on physiological adaptations to resistance training and the underlying mechanisms in skeletal muscle: a narrative review. Front Sports Act Living 3
2. Machado AF et al (2016) Can water temperature and immersion time influence the effect of cold water immersion on muscle soreness? A systematic review and meta-analysis. Sports Med 46:503–514
3. Moore E et al (2022) Impact of cold-water immersion compared with passive recovery following a single bout of strenuous exercise on athletic performance in physically active participants: a systematic review with meta-analysis and meta-regression. Sports Med 52:1667–1688
4. Roberts LA et al (2015) Post-exercise cold water immersion attenuates acute anabolic signalling and long-term adaptations in muscle to strength training. J Physiol 593:4285–4301
5. Higgins TR, Greene DA, Baker MK (2017) Effects of cold water immersion and contrast water therapy for recovery from team sport: a systematic review and meta-analysis. J Strength Cond Res 31:1443–1460
6. Jdidi H, de Bisschop C, Dugué B, Bouzigon R, Douzi W (2024) Optimal duration of whole-body cryostimulation exposure to achieve target skin temperature: influence of body mass index—a randomized cross-over controlled trial. J Physiol Anthropol 43:28
7. Trybulski R et al (2026) Time-dependent microvascular and analgesic responses to cold compression in healthy adults: identifying the optimal cooling duration for recovery. Eur J Appl Physiol. https://doi.org/10.1007/s00421-026-06141-9
8. Trybulski R et al (2024) Optimal duration of cold and heat compression for forearm muscle biomechanics in mixed martial arts athletes: a comparative study. Med Sci Monit 30
9. Wang H, Wang L, Pan Y (2025) Impact of different doses of cold water immersion (duration and temperature variations) on recovery from acute exercise-induced muscle damage: a network meta-analysis. Front Physiol 16
10. Trybulski R et al (2024) The effects of combined contrast heat cold pressure therapy on post-exercise muscle recovery in MMA fighters: a randomized controlled trial. J Hum Kinet 94:127–146
11. Bleakley C et al (2012) Cold-water immersion (cryotherapy) for preventing and treating muscle soreness after exercise. Cochrane Database Syst Rev 2012
12. Casa DJ et al (2007) Validity of devices that assess body temperature during outdoor exercise in the heat. J Athl Train 42:333–342
13. Bongers CCWG, Hopman MTE, Eijsvogels TMH (2015) Using an ingestible telemetric temperature pill to assess gastrointestinal temperature during exercise. J Vis Exp. https://doi.org/10.3791/53258
14. Travers GJS, Nichols DS, Farooq A, Racinais S, Périard JD (2016) Validation of an ingestible temperature data logging and telemetry system during exercise in the heat. Temperature 3:208–219
15. Notley SR, Meade RD, Kenny GP (2021) Time following ingestion does not influence the validity of telemetry pill measurements of core temperature during exercise-heat stress: the journal temperature toolbox. Temperature 8:12–20
16. Huggins R, Glaviano N, Negishi N, Casa DJ, Hertel J (2012) Comparison of rectal and aural core body temperature thermometry in hyperthermic, exercising individuals: a meta-analysis. J Athl Train 47:329–338
17. Morrissey MC, Scarneo-Miller SE, Giersch GEW, Jardine JF, Casa DJ (2021) Assessing the validity of aural thermometry for measuring internal temperature in patients with exertional heat stroke. J Athl Train 56:197–202
18. James CA, Richardson AJ, Watt PW, Maxwell NS (2014) Reliability and validity of skin temperature measurement by telemetry thermistors and a thermal camera during exercise in the heat. J Therm Biol 45:141–149
19. Rossignoli I, Benito PJ, Herrero AJ (2015) Reliability of infrared thermography in skin temperature evaluation of wheelchair users. Spinal Cord 53:243–248

20. Nobre TL et al (2025) The relationship of infrared thermography with the identification of changes in biological markers after EXERCISE: a systematic review. J Bodyw Mov Ther 43:214–220
21. Calvo-Lobo C et al (2023) Intra- and inter-session reliability and repeatability of an infrared thermography device designed for materials to measure skin temperature of the triceps surae muscle tissue of athletes. PeerJ 11:e15011
22. Cheung SS (2015) Responses of the hands and feet to cold exposure. Temperature 2:105–120
23. O'Brien C (2005) Reproducibility of the cold-induced vasodilation response in the human finger. J Appl Physiol 98:1334–1340
24. Huettel M et al (2023) Effects of exercise and cold-water exposure on microvascular muscle perfusion. Ultraschall Med 44:e191–e198
25. Reid G (2005) ThermoTRP channels and cold sensing: what are they really up to? Pflugers Arch 451:250–263
26. Proudfoot CJ et al (2006) Analgesia mediated by the TRPM8 cold receptor in chronic neuropathic pain. Curr Biol 16:1591–1605
27. Algafly AA, George KP (2007) The effect of cryotherapy on nerve conduction velocity, pain threshold and pain tolerance. Br J Sports Med 41:365–369
28. Trybulski R et al (2025) Immediate effect of ice and dry massage during rest breaks on recovery in MMA fighters: a randomized crossover clinical trial study. Sci Rep 15:12323
29. Moore E et al (2023) Effects of cold-water immersion compared with other recovery modalities on athletic performance following acute strenuous exercise in physically active participants: a systematic review, meta-analysis, and meta-regression. Sports Med 53:687–705
30. Costello JT et al (2015) Whole-body cryotherapy (extreme cold air exposure) for preventing and treating muscle soreness after exercise in adults. Cochrane Database Syst Rev 2015
31. Rose C, Edwards K, Siegler J, Graham K, Caillaud C (2017) Whole-body cryotherapy as a recovery technique after exercise: a review of the literature. Int J Sports Med 38:1049–1060
32. Bouzigon R et al (2021) Cryostimulation for post-exercise recovery in athletes: a consensus and position paper. Front Sports Act Living 3
33. Pritchard KA, Saliba SA (2014) Should athletes return to activity after cryotherapy? J Athl Train 49:95–96
34. Cheng JL, MacDonald MJ (2019) Effect of heat stress on vascular outcomes in humans. J Appl Physiol 126:771–781
35. Brunt VE, Howard MJ, Francisco MA, Ely BR, Minson CT (2016) Passive heat therapy improves endothelial function, arterial stiffness and blood pressure in sedentary humans. J Physiol 594:5329–5342
36. Brunt VE, Minson CT (2021) Heat therapy: mechanistic underpinnings and applications to cardiovascular health. J Appl Physiol 130:1684–1704
37. Hussain J, Cohen M (2018) Clinical effects of regular dry sauna bathing: a systematic review. Evid Based Complement Alternat Med 2018
38. Hannuksela ML, Ellahham S (2001) Benefits and risks of sauna bathing. Am J Med 110:118–126
39. Freiwald J et al (2021) A role for superficial heat therapy in the management of non-specific, mild-to-moderate low back pain in current clinical practice: a narrative review. Life 11:780
40. Willis WD (2009) The role of TRPV1 receptors in pain evoked by noxious thermal and chemical stimuli. Exp Brain Res 196:5–11
41. Hiura A (2009) Is thermal nociception only sensed by the capsaicin receptor, TRPV1? Anat Sci Int 84:122–128
42. Ahokas EK, Hennessy RS, Hanstock HG, Kyröläinen H, Ihalainen JK (2025) Effects of post-exercise heat exposure on acute recovery and training-induced performance adaptations: a systematic review. Sports Med Open 11:106
43. Méline T et al (2021) Influence of post-exercise hot-water therapy on adaptations to training over 4 weeks in elite short-track speed skaters. J Exerc Sci Fit 19:134–142
44. Sautillet B et al (2024) Hot water immersion: maintaining core body temperature above 38.5°C mitigates muscle fatigue. Scand J Med Sci Sports 34

45. Ahokas EK, Ihalainen J, Hanstock HG, Savolainen E, Kyröläinen H (2023) A post-exercise infrared sauna session improves recovery of neuromuscular performance and muscle soreness after resistance exercise training. Biol Sport 40:681–689
46. Bieuzen F, Bleakley CM, Costello JT (2013) Contrast water therapy and exercise induced muscle damage: a systematic review and meta-analysis. PLoS One 8:e62356
47. Versey NG, Halson SL, Dawson BT (2012) Effect of contrast water therapy duration on recovery of running performance. Int J Sports Physiol Perform 7:130–140
48. Shattock MJ, Tipton MJ (2012) 'Autonomic conflict': a different way to die during cold water immersion? J Physiol 590:3219–3230
49. Ikäheimo TM (2018) Cardiovascular diseases, cold exposure and exercise. Temperature 5:123–146
50. Romero DJ, Jacobson GP, Roberts RA (2022) The effect of EMG magnitude on the masseter vestibular evoked myogenic potential (mVEMP). J Otol 17:203–210
51. Radtke T et al (2016) Acute effects of Finnish sauna and cold-water immersion on haemodynamic variables and autonomic nervous system activity in patients with heart failure. Eur J Prev Cardiol 23:593–601

Chapter 5
Mechanical Interventions: Compression, Massage, and Lymphatic Drainage

Abstract This chapter examines mechanical recovery interventions through a mechanistic lens, focusing on how externally applied pressure influences tissue fluid balance, venous return, lymphatic transport, and microcirculatory function. It first outlines the physiological basis of external pressure, including revised Starling principles, interstitial fluid dynamics, edema formation, and the role of the lymphatic system in post-exercise recovery. It then analyzes the biological rationale and practical application of sustained compression, intermittent pneumatic compression, massage, and manual lymphatic drainage, highlighting their distinct effects on perfusion, tissue mechanics, inflammatory signaling, soreness, and recovery kinetics. Special attention is given to dose-related factors such as pressure magnitude, stiffness, treatment duration, and anatomical coverage. The chapter also addresses measurement and implementation, defining key variables, summarizing typical values, and discussing validity and reliability of common assessment tools used to quantify pressure, swelling, perfusion, and muscle oxygenation in research and applied sport settings.

5.1 Mechanical Interventions as Pressure-Based Biological Stimuli

Mechanical interventions in recovery should be understood as methods that deliberately alter the pressure environment surrounding muscle, fascia, veins, capillaries, lymphatics, and sensory endings [1]. That framing matters because the revised Starling model indicates that most tissues operate with slight net filtration in the steady state, so excess interstitial fluid is normally cleared predominantly through lymphatic return rather than by sustained venous-side capillary reabsorption [2]. Accordingly, any recovery intervention that changes external pressure may influence recovery not only by improving circulation in a generic sense, but by modifying venous capacitance, interstitial pressure, transcapillary filtration, and the mechanical forces that help propel lymph [3, 4]. This physiology provides a common language for compression, massage, and lymphatic drainage [1].

R. Trybulski, *Biomarker-Guided Physical Recovery Interventions*,
SpringerBriefs in Forensic and Medical Bioinformatics,
https://doi.org/10.1007/978-981-92-0631-5_5

Compression garments and wraps primarily apply sustained external pressure, intermittent pneumatic compression adds cyclical pressure pulses, massage imposes repeated tissue deformation, and manual lymphatic drainage applies lighter directional pressure fluctuations designed to favor fluid mobilization [5]. Although these interventions differ in feel and application, they converge mechanistically on the same core problem in post-exercise tissue: how to control fluid accumulation without worsening nutritive perfusion [6].

A central practical implication is that more pressure is not inherently better pressure [7]. Moderate therapeutic compression can increase pulsatile leg blood flow and improve venous flow characteristics, yet excessive loading in some exercise contexts can raise intramuscular pressure and reduce local tissue oxygenation [8].

Practical Box

In healthy subjects, four-layer leg bandaging applied to an average malleolar pressure of about 40.7 mmHg increased below-knee pulsatile blood flow, showing that external compression can augment rather than suppress arterial inflow under some conditions [8]. In sports contexts, meta-analysis evidence indicates that compression garments enhance venous blood flow more consistently than arterial flow, especially during challenge and recovery phases rather than at quiet rest [6].

Highlights

- *The modern fluid-exchange model makes lymphatic clearance central to recovery physiology because steady-state capillary filtration usually persists even on the venous side.*
- *Mechanical recovery methods should therefore be interpreted through pressure gradients, tissue fluid handling, and microcirculatory consequences rather than by device labels alone.*

Tips

- *When describing any mechanical intervention, report the anatomical region treated, the mode of pressure delivery, session duration, body position, and timing relative to exercise, because these variables materially affect physiological response.*

- *Do not infer delivered pressure from manufacturer sizing alone when the study or practice setting allows direct measurement, because applied pressure varies meaningfully among garments and users.*

Typical Values

- *Compression-garment meta-analysis has commonly grouped studies into pressures below 15 mmHg and at or above 15 mmHg, which is useful for reporting but has not resolved the optimal athlete-recovery dose [9].*
- *In therapeutic bandaging research, a mean malleolar pressure of about 40.7 ± 4.0 mmHg has been reported as sufficient to alter pulsatile leg flow in healthy volunteers [8].*

5.2 Compression-Based Interventions: Garments, Wraps, and Intermittent Pneumatic Compression

Compression garments provide sustained circumferential pressure that may stabilize soft tissue, reduce venous pooling, and modestly improve recovery after exercise [5]. A meta-analysis of 23 studies found small overall benefits for recovery with compression garments, with the clearest effects seen for strength recovery and for recovery after resistance exercise [9]. An earlier meta-analysis of 12 studies also found moderate pooled effects on delayed muscle soreness (DOMS), muscle strength, muscle power, and creatine kinase across 24–72 h after damaging exercise [10]. These findings support compression as a plausible recovery adjunct, but they also show that the magnitude of benefit is usually modest and outcome-specific rather than universally ergogenic [9].

Intermittent pneumatic compression (IPC) should be treated as a distinct intervention because cyclical chamber inflation creates transient hemodynamic surges rather than a static pressure field [11]. In a randomized crossover study in athletes, IPC increased systolic and end-diastolic peak velocities during treatment, with greater effects under the high-pressure protocol, and the hemodynamic changes returned close to baseline within two minutes after treatment ended [12]. A systematic review and meta-analysis of 17 IPC studies in sport concluded that the likely benefits are trivial to small for function, trivial to moderate for pain or soreness, and highly variable for biochemical damage markers [11]. The same review noted that protocols of roughly 20–30 min at around 80 mmHg were the most commonly used athlete-recovery settings [11].

The compression literature also clarifies that pressure dose is only one part of the intervention, and stiffness matters too [13]. The static stiffness index, defined as the pressure difference between standing and lying, helps characterize how a material behaves dynamically during movement, and values below 10 mmHg generally indicate more elastic long-stretch behavior whereas values above 10 mmHg indicate more inelastic short-stretch behavior [13]. That distinction is clinically meaningful because the efficacy of compression depends not only on resting interface pressure but also on how strongly the system resists changes in limb circumference during standing and muscle pump activity [13, 14].

Practical Box

In combat-sport research, recent compression-focused recovery trials have reported that pressure magnitude can shape perfusion and muscle-property responses, which is consistent with the broader idea that athlete recovery depends on delivered dose rather than on the device name alone [15, 16]. In contrast, a crossover study in runners showed that compression stockings during and after a 10-km run increased anterior compartment pressure by about 22 mmHg and reduced tissue oxygenation index by about 11%, illustrating that compression can become counterproductive when local mechanical loading is excessive [7].

Highlights

- *Compression garments have the strongest evidence for modest benefits in soreness and some recovery metrics, not for large direct gains in immediate performance.*
- *IPC appears most defensible as a short post-exercise recovery tool for soreness management and transient flow enhancement rather than as a universal anti-damage intervention.*

Tips

- *For garments, report whether pressure was directly measured at the skin-device interface or only estimated from garment class or manufacturer sizing, because pressure heterogeneity is a recurrent limitation in the literature.*

- *For IPC, report chamber sequence, inflation pressure, cycle timing, total duration, limb position, and the elapsed time between exercise cessation and treatment onset, because these variables condition the acute hemodynamic response.*

Typical Values

- *Athlete IPC protocols are most commonly reported at about 20–30 min and around 80 mmHg, but these values should be presented as common research settings rather than definitive universal prescriptions [11].*
- *One sports-compression trial in rugby used suits delivering 13–31 mmHg, and compression-garment meta-analysis has stratified studies around the 15 mmHg threshold [17].*

5.3 Massage as a Mechanical Signaling Intervention

Massage should be presented not simply as a comfort intervention but as a patterned mechanical deformation of tissue [18]. In healthy participants, massage applied after fatiguing isometric lumbar exercise altered skin and intramuscular circulatory responses, supporting the idea that manual deformation can change local hemodynamics in fatigued tissue [18]. That circulatory effect is biologically plausible because repeated compression, shear, and release can transiently modify vessel caliber, interstitial pressure, and local sensory input [19].

Massage may also act below the macroscopic circulatory level [20]. After exercise-induced muscle damage, massage activated focal adhesion kinase and ERK1/2 signaling, potentiated PGC-1alpha-related mitochondrial biogenesis signaling, and mitigated the rise in NF-kB nuclear accumulation, which suggests that mechanical loading can influence post-exercise inflammatory biology and cellular adaptation pathways [20]. This makes massage relevant to an evidence-based recovery text because it provides one of the clearest examples in the literature of a manual intervention producing measurable intracellular responses rather than only subjective relief [20].

At the outcome level, however, massage should be described with appropriate restraint [21]. A systematic review and meta-analysis of 29 studies found no evidence that sports massage improves strength, jump, sprint, endurance, or fatigue, although it did show small but significant improvements in flexibility and DOMS [21]. Another meta-analysis found that massage after strenuous exercise reduced DOMS, improved some force outcomes, and lowered creatine kinase (CK), but the overall evidence still supports soreness-related benefits more consistently than robust performance enhancement [22]. The practical message is that massage is best framed as a recovery-modulating intervention with sensory, circulatory, and signaling effects, not as a substitute for training adaptation or a guaranteed performance booster [21].

Practical Box

In combat-sport athletes exposed to eccentric plyometric loading, recent work found that intense sports massage increased perfusion, whereas lymphatic drainage in the same trial had stronger effects on pain and inflammatory markers, which is clinically useful when choosing between deeper and lighter manual strategies [23]. In an earlier DOMS trial, massage administered two hours after exercise-induced muscle injury reduced soreness at 48 h but did not improve hamstring function, which fits the broader conclusion that symptom relief is more reliable than performance restoration [24].

Highlights

- *Massage has plausible local circulatory and cellular signaling effects, but the strongest practical evidence remains for DOMS reduction and small flexibility benefits.*
- *Avoid overstating massage as an ergogenic tool and instead position it as a targeted intervention for symptom burden and selected biological responses.*

Tips

- *Standardize body region, stroke type, session duration, timing after exercise, therapist qualification, and the co-interventions used, because massage meta-analyses are limited by major protocol heterogeneity.*
- *Match outcome measures to the mechanism being targeted, using soreness, pressure-pain threshold, local perfusion, and short-horizon muscle function rather than expecting all massage protocols to change every recovery endpoint.*

Typical Values

- *Mechanistic studies have used brief local massage exposures of about 5 min, whereas athlete recovery trials have also used repeated 30-min sessions over 72 h, so duration should be reported as a meaningful dose variable [18].*
- *The current evidence base suggests that the expected effect size is larger for DOMS and flexibility than for sprint, jump, or endurance performance [21].*

5.4 Manual Lymphatic Drainage

Manual lymphatic drainage (MLD) should be separated from massage because its theoretical target is different [25]. Lymph flow depends on both intrinsic pumping by collecting lymphatics and extrinsic tissue forces, and effective propulsion requires coordinated lymphangion contraction and competent valves [3]. Because initial lymphatics and precollectors are structurally adapted to fluid uptake rather than to resist deep compressive loading, MLD is conceptually a low-load, rhythmical, directional intervention designed to modulate interstitial pressures and facilitate fluid entry into the lymphatic system [4].

This distinction is especially relevant in the post-exercise context when swelling, heaviness, or delayed fluid clearance appears more prominent than mechanical muscle tightness alone [26]. A review focused on orthopedic injury management described MLD as a manual therapy that assists lymphatic function by promoting variations in interstitial pressure through light hand movements [25]. That wording is compatible with revised microvascular physiology, because once excess interstitial fluid is present, the lymphatic system becomes the principal clearance route, and mechanical techniques that favor lymph uptake are theoretically attractive [2].

The exercise-recovery evidence for MLD is promising but still smaller than the literatures for compression garments or generic massage [26, 27]. In recreational athletes, MLD after treadmill exercise was associated with a faster decline in lactate dehydrogenase (LDH) and aspartate aminotransferase (AST) over recovery, with CK showing a similar but non-significant pattern [26]. In mixed martial arts (MMA) athletes, physical methods of lymphatic drainage improved post-exercise regeneration markers including maximal strength, muscle tension, pain threshold, lactate, and creatine kinase over the short recovery period [27]. More recently, in combat-sport athletes recovering from eccentric plyometric loading, lymphatic drainage showed stronger effects on pain, inflammation, and some muscle-damage markers than intense sports massage, whereas the massage condition more consistently increased perfusion [23].

The correct position is that MLD is most persuasive when the clinical picture includes edema-like fluid retention, pressure sensitivity, or a need for lower-load recovery work [25]. It should also be stated clearly that, in edema-dominant conditions, MLD usually belongs within a broader compression-oriented strategy rather than replacing compression entirely [28].

Practical Box

In a randomized treadmill study, seven treated athletes received MLD after exercise and showed significantly lower LDH at 48 h than controls, supporting a biologically plausible effect on recovery kinetics [26]. In combat-sport research, a randomized trial applied six 30-min recovery sessions over 72 h and found that lymphatic drainage and intense massage both aided recovery, but with different response profiles, which supports a phenotype-based rather than modality-loyal clinical approach [23].

Highlights

- *MLD is not simply "gentle massage," because its rationale is tied to lymphatic uptake and fluid mobilization rather than to deep tissue deformation alone.*
- *The current sports evidence is encouraging but still smaller than the compression and massage literatures, so claims should remain targeted and not universalized.*

Tips

- *Session duration and frequency are reported more consistently than therapist-applied pressure magnitude, so practice guidelines should standardize patient position, treated regions, sequence, duration, and outcome timing rather than claim a universal mmHg dose for MLD.*
- *When swelling is a major target, pair MLD with an external containment strategy or at least document whether compression was also used, because edema-focused evidence frequently treats compression as foundational care.*

Typical Values

- *Athlete MLD studies have reported single post-exercise treatment sessions and repeated 30-min sessions delivered across 72 h, so frequency and total exposure should be presented as part of dose.*
- *The literature supports light pressure with different hand movements as a definitional feature of MLD, but it does not yet provide a well-validated universal pressure target for sports recovery practice.*

5.5 Measurement, Validity, and Implementation

For compression, the primary mechanical variable is interface pressure, usually expressed in mmHg, and an essential complementary variable is stiffness, because the same resting pressure can behave differently during posture change and movement [13]. The Kikuhime probe has good accuracy and precision for in vivo assessment, and interface-pressure devices such as Pliance X and newer portable sensors have shown clinically acceptable to excellent reliability in validation studies [13].

For edema or fluid status, the most practical volume measures remain tape or girth measurement, water displacement, and perometry, while bioimpedance spectroscopy (BIS) and tissue dielectric constant (TDC) add information about extracellular fluid and local tissue water rather than gross volume alone [29–32]. A systematic review found good reliability and validity for BIS, water volumetry, tape measurement, and perometry, with BIS being particularly useful for extracellular-fluid change and the volume methods being more established for larger stage-related limb changes [29]. In post-surgical swelling work, BIS showed intra- and inter-evaluator intraclass correlation (ICCs) from 0.89 to 0.99 and high responsiveness, which supports its use when the goal is to capture subtle fluid changes rather than only circumferential enlargement [31]. For local superficial tissue water, TDC has high interobserver agreement at the ankle and lower leg, with ICCs of 0.94 at both sites, but the foot is less stable and should be interpreted more cautiously [32].

For perfusion and oxygenation, laser Doppler flowmetry (LDF), laser speckle coherence imaging (LSCI), and near-infrared spectroscopy (NIRS) serve different purposes and should not be treated as interchangeable [33–35]. LDF is useful for cutaneous microcirculation and acute vasodilatory responses, but baseline repeatability can be weak and improves substantially when protocols use standardized provocations such as local heating or occlusion [33]. LSCI offers non-contact perfusion mapping and has shown good reproducibility in some microvascular applications, which makes it attractive when spatial heterogeneity matters [35]. NIRS is the preferred field-friendly option for muscle oxygenation, and recent reliability work reported tissue-oxygenation-index coefficients of variation of 2.7–10.2% with ICCs around 0.86–0.90 in squat-exercise testing [34].

Practical Box

In MMA athlete recovery studies, investigators have already used combined outcome batteries including maximal force, muscle tension, pain threshold, lactate, creatine kinase, and local perfusion, which is close to the kind of multidomain framework this book advocates [27]. The key translational lesson is that no single endpoint captures "recovery," so mechanical interventions should be judged against the dominant problem to solve, such as soreness, swelling, venous pooling, impaired reoxygenation, or altered muscle mechanical properties [36].

Highlights

- *Measure the dose actually delivered, then measure the biological response actually targeted.*
- *For fluid-related endpoints, tape, water displacement, perometry, BIS, and TDC each answer slightly different questions, so selection should follow mechanism and feasibility rather than habit.*

Tips

- *For interface pressure, record after a short stabilization period and document posture, because standing and lying values are both needed when SSI is part of the analysis.*
- *For volume or tissue-water measurements, mark anatomical landmarks, repeat measures in duplicate, and keep time-of-day, hydration state, and prior activity as constant as possible, because observer and condition effects remain non-trivial even with validated tools.*
- *For LDF and NIRS, fix probe placement carefully and use the same provocation and environmental conditions across sessions, because reliability improves when acquisition is standardized.*

Typical Values

- *SSI values below 10 mmHg generally indicate more elastic long-stretch behavior, whereas values above 10 mmHg indicate more inelastic short-stretch behavior [13].*
- *TDC upper normal reference limits reported for women were 35.2 at the ankle and 38.3 at the lower leg, and the effective measurement depth in that study was 2.5 mm [32].*
- *BIS lower-limb swelling studies have reported intra- and inter-evaluator ICCs from 0.89 to 0.99, while NIRS tissue-oxygenation-index testing has shown coefficients of variation of 2.7–10.2% with ICCs around 0.86–0.90 [31].*

References

1. Partsch H (2012) Compression therapy: clinical and experimental evidence. Ann Vasc Dis 5:416–422
2. Levick JR, Michel CC (2010) Microvascular fluid exchange and the revised Starling principle. Cardiovasc Res 87:198–210
3. Scallan JP, Zawieja SD, Castorena-Gonzalez JA, Davis MJ (2016) Lymphatic pumping: mechanics, mechanisms and malfunction. J Physiol 594:5749–5768
4. Breslin JW (2014) Mechanical forces and lymphatic transport. Microvasc Res 96:46–54
5. MacRae BA, Cotter JD, Laing RM (2011) Compression garments and exercise. Sports Med 41:815–843
6. O'Riordan SF, Bishop DJ, Halson SL, Broatch JR (2023) Do sports compression garments alter measures of peripheral blood flow? A systematic review with meta-analysis. Sports Med 53:481–501
7. Rennerfelt K, Lindorsson S, Brisby H, Baranto A, Zhang Q (2019) Effects of exercise compression stockings on anterior muscle compartment pressure and oxygenation during running: a randomized crossover trial conducted in healthy recreational runners. Sports Med 49:1465–1473
8. Mayrovitz HN, Larsen PB (1997) Effects of compression bandaging on leg pulsatile blood flow. Clin Physiol 17:105–117
9. Brown F et al (2017) Compression garments and recovery from exercise: a meta-analysis. Sports Med 47:2245–2267
10. Hill J, Howatson G, van Someren K, Leeder J, Pedlar C (2014) Compression garments and recovery from exercise-induced muscle damage: a meta-analysis. Br J Sports Med 48:1340–1346
11. Maia F, Nakamura FY, Sarmento H, Marcelino R, Ribeiro J (2024) Effects of lower-limb intermittent pneumatic compression on sports recovery: a systematic review and meta-analysis. Biol Sport 41:263–275
12. Maia F, Machado MVB, Silva G, Nakamura FY, Ribeiro J (2024) Hemodynamic effects of intermittent pneumatic compression on athletes: a double-blinded randomized crossover study. Int J Sports Physiol Perform 19:932–938
13. Partsch H (2006) The static stiffness index: a simple method to assess the elastic property of compression material in vivo. Dermatologic Surg 31:625–630

14. Mosti GB, Mattaliano V (2007) Simultaneous changes of leg circumference and interface pressure under different compression bandages. Eur J Vasc Endovasc Surg 33:476–482
15. Trybulski R et al (2025) Effect of pneumatic and cold compression on muscle performance and recovery in combat sports athletes. Sci Rep 15:44993
16. Trybulski R et al (2026) Postexercise physical recovery methods for combat sports: a scoping review. Int J Sports Med 47:3–23
17. McMaster DT, Beaven CM, Mayo B, Gill N, Hébert-Losier K (2017) The efficacy of wrestling-style compression suits to improve maximum isometric force and movement velocity in well-trained male rugby athletes. Front Physiol 8
18. Mori H et al (2004) Effect of massage on blood flow and muscle fatigue following isometric lumbar exercise. Med Sci Monit 10:CR173-8
19. Bliss MR (1998) Hyperaemia. J Tissue Viabil 8:4–13
20. Crane JD et al (2012) Massage therapy attenuates inflammatory signaling after exercise-induced muscle damage. Sci Transl Med 4
21. Davis HL, Alabed S, Chico TJA (2020) Effect of sports massage on performance and recovery: a systematic review and meta-analysis. BMJ Open Sport Exerc Med 6:e000614
22. Guo J et al (2017) Massage alleviates delayed onset muscle soreness after strenuous exercise: a systematic review and meta-analysis. Front Physiol 8
23. Trybulski R et al (2025) Effects of lymphatic drainage and intense sports massage on muscle properties, damage, and function after eccentric plyometric training: a randomized controlled parallel trial. Eur J Appl Physiol. https://doi.org/10.1007/s00421-025-06101-9
24. Hilbert JE, Sforzo GA, Swensen T (2003) The effects of massage on delayed onset muscle soreness. Br J Sports Med 37:72–75
25. Majewski-Schrage T, Snyder K (2016) The effectiveness of manual lymphatic drainage in patients with orthopedic injuries. J Sport Rehabil 25:91–97
26. Schillinger A et al (2006) Effect of manual lymph drainage on the course of serum levels of muscle enzymes after treadmill exercise. Am J Phys Med Rehabil 85:516–520
27. Zebrowska A, Trybulski R, Roczniok R, Marcol W (2019) Effect of physical methods of lymphatic drainage on postexercise recovery of mixed martial arts athletes. Clin J Sport Med 29:49–56
28. McNeely ML et al (2004) The addition of manual lymph drainage to compression therapy for breast cancer related lymphedema: a randomized controlled trial. Breast Cancer Res Treat 86:95–106
29. Hidding JT et al (2016) Measurement properties of instruments for measuring of lymphedema: systematic review. Phys Ther 96:1965–1981
30. Brodovicz KG et al (2009) Reliability and feasibility of methods to quantitatively assess peripheral edema. Clin Med Res 7:21–31
31. Pichonnaz C, Bassin J-P, Lécureux E, Currat D, Jolles BM (2015) Bioimpedance spectroscopy for swelling evaluation following total knee arthroplasty: a validation study. BMC Musculoskelet Disord 16:100
32. Jensen MR, Birkballe S, Nørregaard S, Karlsmark T (2012) Validity and interobserver agreement of lower extremity local tissue water measurements in healthy women using tissue dielectric constant. Clin Physiol Funct Imaging 32:317–322
33. Eun HC (1995) Evaluation of skin blood flow by laser Doppler flowmetry. Clin Dermatol 13:337–347
34. Corral-Pérez J et al (2024) Reliability of near-infrared spectroscopy in measuring muscle oxygenation during squat exercise. J Sci Med Sport 27:805–813
35. Mahé G, Humeau-Heurtier A, Durand S, Leftheriotis G, Abraham P (2012) Assessment of skin microvascular function and dysfunction with laser speckle contrast imaging. Circ Cardiovasc Imaging 5:155–163
36. Dupuy O, Douzi W, Theurot D, Bosquet L, Dugué B (2018) An evidence-based approach for choosing post-exercise recovery techniques to reduce markers of muscle damage, soreness, fatigue, and inflammation: a systematic review with meta-analysis. Front Physiol 9

Chapter 6
Needling-Based and Neuromodulatory Interventions

Abstract This chapter examines needling-based and neuromodulatory interventions within an evidence-based recovery framework, with emphasis on dry needling as a modulator of vascular, neuromechanical, and nociceptive responses. It outlines the biological rationale for targeting myofascial trigger points and related soft-tissue dysfunction, and explains how needling may influence local perfusion, tissue oxygenation, muscle stiffness, pain sensitivity, and sensorimotor behavior. The chapter also describes the main tools used to quantify these responses, including algometry, myotonometry, elastography, thermography, near-infrared spectroscopy, and Doppler-based approaches, with attention to validity, reliability, and implementation standards. Practical sections address dose parameters, procedural considerations, athlete-specific application, safety, adverse events, and the cautious integration of electrical augmentation strategies. Throughout, the chapter connects physiological mechanisms with clinically meaningful measurement, helping readers interpret when, how, and why needling interventions may be relevant in recovery-oriented practice.

6.1 Why Needling Belongs in a Biomarker-Guided Recovery Framework

Dry needling is most consistently described as the insertion of a solid filiform needle into muscle or related soft tissue with the aim of modulating a neuromusculoskeletal dysfunction, most commonly a myofascial trigger point [1]. Within the logic of this book, dry needling is best treated not as an isolated pain-relief technique but as a short-latency biological stimulus that can alter local perfusion, muscle mechanical behavior, and nociceptive processing in the same tissue unit [2]. Umbrella and meta-analytic evidence suggests that dry needling is generally superior to sham dry needling for short-term pain reduction, while functional gains are smaller, more heterogeneous, and less securely established over longer follow-up periods [3].

R. Trybulski, *Biomarker-Guided Physical Recovery Interventions*,
SpringerBriefs in Forensic and Medical Bioinformatics,
https://doi.org/10.1007/978-981-92-0631-5_6

Practical Box

In athletes, dry needling is usually considered when a small number of high-value loci combine elevated pain sensitivity with palpable stiffness or tone that cannot be left to normalize passively within the available training microcycle [4, 5]. *Pilot and randomized athlete studies show that biological changes can be captured in MMA athletes, triathletes, and ballet dancers, but they also show that the immediate post-needling window is not always the optimal window for judging usefulness* [4, 6, 7].

Highlights

- *The strongest current signal for dry needling is short-term modulation of pain sensitivity, with additional but less uniform effects on perfusion and mechanical behavior* [3].

6.2 Biological Substrate of the Needled Muscle Region

The pathophysiological substrate most often targeted by dry needling is the myofascial trigger point, which has been described as a hyperirritable locus within a taut band that contains both a sensory component and an active locus associated with spontaneous electrical activity [8]. Current trigger-point models propose that spontaneous electrical activity arises from dysfunctional endplates and that sustained local contracture may create a microenvironment of relative ischemia, nociceptor sensitization, and altered motor control [2, 9]. Microdialysis work has shown that active trigger points contain elevated concentrations of biochemicals associated with pain and inflammation, supporting the idea that the target tissue is biologically altered rather than merely tender on palpation [10]. Experimental needling studies further show that when local twitch responses are elicited, spontaneous electrical activity can be reduced, which provides one plausible mechanism for short-term modulation of the trigger-point state [11]. From a recovery perspective, the practical implication is that dry needling is not acting on "pain" alone but on a coupled system of endplate dysfunction, nociceptive chemistry, motor guarding, and local vascular behavior [10].

Practical Box

When an athlete presents with a focal painful band after repeated eccentric loading, the relevant question is not only whether the site hurts, but whether it is also mechanically altered and likely to be driving protective motor behavior

[9]. *This is one reason dry needling fits your book better than a purely symptom-based manual, because it can be interpreted through concurrent changes in tissue mechanics, perfusion, and pressure pain threshold (PPT)* [2].

Highlights

- *The needled locus is biologically active tissue, not just a painful spot* [10].
- *The most defensible mechanism model combines dysfunctional endplates, altered chemistry, nociceptor sensitization, and downstream central amplification* [8].

6.3 Dry Needling as a Vascular Intervention

Human experimental work in the upper trapezius has shown that a single dry needling application increases blood flow and oxygen saturation at the treated point, with values remaining elevated for at least 15 min after needle removal, while distant sites change only minimally [12]. Infrared thermography work in gluteus minimus-related sciatica has shown short-term vasodilation and temperature increases in the referred pain area after dry needling of active trigger points, which supports the view that autonomic and vasomotor responses can extend beyond the exact puncture point [13]. These data justify describing dry needling as a local vascular perturbation with possible segmental autonomic expression, but they do not justify the stronger claim that dry needling reliably improves whole-limb or long-duration recovery perfusion in every context [2].

6.3.1 Perfusion Measures and how to Interpret them

Laser Doppler flowmetry quantifies microvascular flux in perfusion units and is useful for continuous local monitoring, but its absolute reliability is only moderate unless the protocol is stabilized carefully [14]. Near-infrared spectroscopy is a noninvasive method that tracks the oxygenation state of hemoglobin and myoglobin and is therefore useful when the aim is to describe the balance between oxygen delivery and oxygen utilization [15]. Infrared thermography measures skin-surface temperature rather than intramuscular blood flow, so it should be interpreted as an indirect vasomotor marker that requires strict environmental standardization [16]. Doppler ultrasound is preferable when conduit-artery flow is the outcome of interest and is recommended for resting or light-to-moderate exercise conditions when handled by a skilled operator [17].

Practical Box

In triathletes with latent triceps surae trigger points, investigators have already combined thermography with PPT to test whether needling changed both pain sensitivity and superficial thermal behavior in the immediate window [6]. *In mixed martial arts (MMA) athletes, recent randomized studies have extended this logic by pairing perfusion outcomes with mechanical and force variables, which is closer to the translational standard your book should advocate* [4].

Highlights

- *Dry needling can acutely increase local blood flow and oxygenation* [12].
- *Skin temperature is useful, but it is an indirect vascular signal and must not be misread as direct intramuscular perfusion* [16].

Tips

- *Use exactly the same probe location, posture, and stabilization period for repeated laser Doppler flowmetry (LDF) or near-infrared spectroscopy (NIRS) measurements, because reliability improves once the physiological signal has stabilized.*
- *For thermography, adopt a standardized checklist for room setup, participant preparation, camera placement, and image analysis rather than informal image capture.*
- *When perfusion is the endpoint, collect baseline, immediate post-needling, and at least one delayed follow-up measurement rather than relying on a single post-treatment snapshot.*

Typical Values

- *In upper trapezius, blood flow and oxygen saturation remained elevated for at least 15 min after needling at the treated point* [12].

- *In active gluteus minimus trigger points associated with sciatica, referred-pain area temperature increases of roughly 1.2–1.6 °C in the thigh and about 0.4 °C in the calf were reported after dry needling, illustrating that thermographic responses can be biologically visible but context-specific* [13].

6.4 Dry Needling as a Neuromechanical Intervention

Several randomized and controlled studies indicate that dry needling can reduce muscle stiffness and sometimes tone in upper trapezius, soleus, and gastrocnemius regions, although the magnitude and persistence of those effects vary across muscles and studies [18]. A plausible mechanistic reading is that needling perturbs the local contracture-endplate-nociception loop, which then reduces protective muscle guarding and changes the passive mechanical behavior recorded by instrumented devices [11]. At the same time, not every study finds a between-group stiffness reduction exactly at the trigger point on elastography, which means the observed response depends materially on device physics, region-of-interest placement, tissue depth, and the timing of reassessment [19].

6.4.1 Measures of Tone, Stiffness, and Elasticity

MyotonPRO quantifies oscillation frequency, dynamic stiffness, and decrement by delivering a brief mechanical impulse to the tissue and has shown generally high intra-rater and inter-rater reliability across multiple muscles [20, 21]. Ultrasound elastography quantifies muscle hardness or shear-related stiffness and has demonstrated acceptable reliability and validity, but operator technique and region selection remain decisive for data quality [22]. Myotonometry and elastography should therefore be treated as complementary rather than interchangeable measures, because they probe different depths and different mechanical constructs [23].

Practical Box

In healthy volunteers with latent upper-trapezius trigger points, one randomized trial showed lower dynamic stiffness at 72 h and lower tone at 72 h after dry needling compared with sham [24]. *In MMA athletes and other sport cohorts, the strongest practical value of these measures is not diagnostic labeling but monitoring whether a needled muscle is trending toward lower stiffness and higher pain tolerance over the next 24–72 h* [4, 5].

Highlights

- *Dry needling can change instrumented muscle mechanics, but the response is not uniform across muscles, devices, or timepoints.*
- *Neuromechanical outcomes become much more interpretable when combined with perfusion and PPT outcomes from the same session.*

Tips

- *Mark the exact test point before intervention and maintain the same posture and tissue relaxation state during repeated measures.*
- *Avoid comparing values collected under different recent loading states, because exercise history can influence both perfusion and muscle mechanics.*
- *Report the device, parameter, unit, number of repetitions, and averaging rule, because poor dose and measurement reporting is a major weakness of the literature* [25].

Typical Values

- *In one upper-trapezius randomized trial, dynamic stiffness decreased from 278.74 ± 38.40 to 261.54 ± 33.64 N/m and tone decreased from 16.62 ± 1.27 to 15.88 ± 1.31 Hz by 72 h after dry needling* [24]
- *In test-retest work on myotonometry, intraclass correlation (ICC) values for mechanical parameters commonly fall in the good-to-excellent range, but muscle-specific minimal detectable changes (MDC) should be treated as site-dependent rather than universal* [16, 20].

6.5 Dry Needling as a Neuromechanical Intervention

Pain modulation is the most consistent clinical signal associated with dry needling, with meta-analyses showing benefits from the immediate post-treatment window through several weeks in many musculoskeletal contexts [26]. Systematic review data also show that dry needling tends to increase pressure pain threshold in the immediate to 12-week window, which makes PPT a particularly useful mechanistic bridge between symptom relief and quantifiable sensory change [27]. Some trials have reported increases in widespread pressure pain threshold after trigger-point dry needling, which suggests that at least part of the effect is not purely local and may involve reduced central excitability [28, 29]. A pain-neuroscience interpretation is therefore justified, but it must remain modest since the best-supported effects are short-term, and not every patient demonstrates the same spread, duration, or functional consequence of hypoalgesia [29]. Immediate interpretation also requires caution because post-needling soreness is common and can temporarily obscure whether the intervention is beneficial or simply irritating [30].

6.5.1 The Local Twitch Response

The local twitch response remains one of the most debated features of dry needling, because some experimental work links it to suppression of spontaneous electrical activity and some clinical work links it to immediate mechanical or functional change [11, 31]. At the same time, contemporary reviews conclude that eliciting a twitch is not universally necessary for positive clinical effects and that overly aggressive twitch-oriented dosing may increase soreness without guaranteeing superior outcomes [32, 33]. Thus, twitch response may be a biological signal of target engagement in some contexts, but it should not be elevated to a universal marker of adequate dosage [33].

Practical Box

In combat-sport athletes, dry needling has been associated with improved PPT alongside changes in perfusion and biomechanical measures, which supports the practical use of multimarker reassessment instead of a pain score alone [4, 5]. *In acute mechanical neck pain, trigger-point dry needling has also been shown to increase widespread PPT and cervical range of motion (ROM), illustrating that a local intervention can produce regionally broader sensory change* [28].

Highlights

- *Hypoalgesia is the most reproducible clinical output of dry needling* [26].
- *The local twitch response is mechanistically interesting but not a mandatory end point* [32].

6.6 Measuring the Response to Dry Needling

Pressure pain threshold (PPT) is the most useful quantitative nociceptive endpoint for this chapter because it operationalizes the minimum pressure that becomes painful and has acceptable reliability in both clinical and research settings [34, 35]. Visual Analogue Scale (VAS) and Numeric Rating Scale (NRS) remain acceptable unidimensional pain tools, but current evidence does not clearly establish one as uniformly superior across musculoskeletal conditions, so consistency of use matters more than brand preference [36]. A biomarker-guided design should therefore pair one patient-reported measure with at least one objective biological measure from perfusion or muscle mechanics, because subjective analgesia alone cannot distinguish local mechanical normalization from temporary desensitization [29].

Practical Box

The strongest athlete-facing study designs already pair PPT with force, perfusion, and mechanical measures, which is why they are more informative than trials that report only pain [4]. *This same framework can be transferred to rehabilitation clinics by using algometry plus either MyotonPRO or elastography and one accessible vascular measure* [20].

Highlights

- *We should not treat pain ratings as sufficient endpoints* [29].
- *PPT is the most useful bridge variable between symptom change and mechanism* [27].

Tips

- *For PPT, use the same marked point, perpendicular application, and trained assessor across sessions, because reliability depends strongly on procedural consistency.*
- *Report intraclass correlation (ICC), standard error of the mean (SEM), and minimal detectable changes (MDC) whenever possible, because a treatment effect smaller than measurement error is biologically ambiguous.*
- *Use the same follow-up windows across all endpoints, because pain, perfusion, and stiffness do not necessarily peak at the same time.*

Typical Values

- *In chronic neck pain with trapezius trigger points, PPT showed excellent intra- and inter-rater reliability with ICC values of 0.752–0.874, SEM values of 0.18–0.22 kg/cm^2, and MDC values of 0.45–0.62 kg/cm^2* [35].
- *In a broader neck-pain reliability study, intrarater ICC values of 0.94–0.97 and interrater ICC values of 0.79–0.90 were reported for digital algometry, which supports routine quantitative use after basic training* [34].

6.7 Dosing and Implementation in Athletes

One of the clearest weaknesses in the dry needling literature is inadequate standardization of dose, including insertion depth, number of insertions, manipulation style, retention time, treated loci, and adverse-event reporting [25]. That matters especially in athlete recovery, where the practical question is not merely whether dry needling "works" but whether a specific dose improves the desired endpoint within the available recovery window [4].

Current athlete data suggest that dry needling should usually be treated as a recovery-window or between-session intervention rather than as a same-hour performance enhancer, because immediate soreness may coexist with more favorable biological changes over the following 24–72 h [7, 37]. In practice, target selection should prioritize muscles or subregions that are simultaneously pain-sensitive, mechanically altered, and functionally relevant to the next load exposure, because broad multi-site needling without a biomarker rationale is hard to justify in a short recovery cycle [29].

These considerations are further supported by findings identified in a recent systematic review with an evidence gap map of dry needling in sports and sport recovery [38]. The analysis, encompassing 24 studies and 580 athletes, revealed substantial heterogeneity in the reporting of dosage parameters, including insertion depth, number of insertions, retention time, and procedural characteristics, as well as inconsistent documentation of adverse events. Moreover, the majority of studies focused primarily on short-term pain outcomes, with comparatively limited attention given to objective biomechanical, neuromuscular, or performance-related variables [38]. The identified research gaps underscore the necessity of adopting a more structured, biomarker-guided framework when planning and interpreting dry needling interventions in athletic populations.

6.7.1 Electrical Augmentation: PENS, IMES, DNES

Electrical augmentation after or through needling is promising but still supported mainly by small and heterogeneous trials [39, 40]. Randomized neck-pain studies suggest that percutaneous electrical nerve stimulation (PENS) can improve some short-term outcomes such as PPT, disability, or post-needling soreness relative to dry needling alone in selected contexts [41]. However, frequency comparisons do not consistently show superiority of one stimulation frequency over another, and current evidence is not strong enough to promote electrical augmentation as a default extension for all athletes [39, 42].

Practical Box

In mixed martial arts athletes, biomarker changes after dry needling have been reported up to 48 h, which fits a between-sessions recovery model better than a pre-bout model [5]. *In chronic neck pain, combining PENS with dry needling has reduced post-needling soreness more than dry needling alone, which suggests one possible strategy when soreness management is clinically important* [30].

Highlights

- *Dose matters, but the literature still reports dose poorly.*
- *In athletes, the most defensible scheduling is usually 24–72 h before the next key performance exposure until the individual response is known.*

Tips

- *Treat the first exposure as a profiling session and reassess soreness, PPT, and at least one performance-relevant variable before using dry needling close to competition.*
- *Avoid aggressive multi-site dosing immediately before activities requiring explosive output or fine motor precision until the athlete's individual response curve is known.*
- *When using PENS or intramuscular electrical stimulation (IMES), document frequency, intensity, duration, and whether stimulation was added after DN or delivered intramuscularly through the needle pathway.*

Typical Values

- *In one upper-trapezius randomized trial, post-needling soreness decreased by about 33% at 30 min, about 81% at 24 h, and had resolved by 72 h* [37].
- *In another chronic neck-pain trial, dry needling plus PENS reduced immediate neck pain intensity and short-term soreness more than dry needling alone* [30].

6.8 Safety, Contraindications, and Adverse-Event Reporting

Large survey data indicate that minor adverse events such as mild bleeding, bruising, and pain during treatment are common, whereas major adverse events are rare [43]. This does not make dry needling trivial, because risk is highly anatomy-dependent and preventable complications remain possible in thoracic, cervical, and deep neurovascular regions [44].

The minimum safety standard should include informed consent, region-specific anatomical reasoning, sterile technique, documentation of immediate reactions, and explicit scheduling that accounts for expected soreness [45]. Because RCTs still underreport adverse events and dosing parameters, this chapter should explicitly call for standardized documentation of event type, severity, site, onset, duration, and effect on subsequent training [25].

In addition to local and anatomical risks, vasovagal responses represent a clinically relevant autonomic safety consideration in dry needling practice. Recent controlled observational research has demonstrated that vasovagal episodes following dry needling are associated with measurable physiological changes, including alterations in heart rate, heart rate variability, and pain sensitivity parameters, indicating that these reactions reflect a coordinated autonomic response rather than a purely psychogenic event [46]. This finding is particularly important in athletic populations, where interventions are often performed in states of fatigue, dehydration, high sympathetic activation, or post-exertional hemodynamic fluctuation. Under such conditions, susceptibility to transient hypotension, dizziness, or presyncope may be increased. Therefore, clinicians should consider pre-procedural screening for prior vasovagal history, ensure appropriate positioning and supervision during and immediately after needling, and avoid aggressive multi-site dosing in physiologically stressed athletes.

Practical Box

In athlete environments, even a "minor" event such as localized bruising or soreness can alter the next technical or strength session if the intervention was scheduled poorly. That is why safety in sport is not only about avoiding rare major events, but also about predicting the training consequences of common minor ones.

Highlights

- *Minor adverse events are common, major ones are rare, and both must be reported.*
- *Safety is a function of anatomy, dose, and scheduling, not merely of needle size.*

Typical Values

- *In one large survey, 36.7% of reported therapeutic dry needling treatments were associated with minor adverse events, while major events occurred at a rate below 0.1% and about 1 per 1024 treatments* [44].
- *n another survey, 19.18% of treatments were associated with mild adverse events and no significant adverse events were reported* [43].

References

1. Dunning J et al (2014) Dry needling: a literature review with implications for clinical practice guidelines. Phys Ther Rev 19:252–265
2. Cagnie B et al (2013) Physiologic effects of dry needling. Curr Pain Headache Rep 17:348
3. Chys M et al (2023) Clinical effectiveness of dry needling in patients with musculoskeletal pain—an umbrella review. J Clin Med 12:1205
4. Trybulski R, Stanula A, Żebrowska A, Podleśny M, Hall B (2024) Acute effects of the dry needling session on gastrocnemius muscle biomechanical properties, and perfusion with latent trigger points—a single-blind randomized controlled trial in mixed martial arts athletes. J Sports Sci Med 136–146. https://doi.org/10.52082/jssm.2024.136
5. Trybulski R et al (2024) Biomechanical profile after dry needling in mixed martial arts. Int J Sports Med. https://doi.org/10.1055/a-2342-3679
6. Benito-de-Pedro et al (2019) Effectiveness between dry needling and ischemic compression in the triceps surae latent myofascial trigger points of triathletes on pressure pain threshold and thermography: a single blinded randomized clinical trial. J Clin Med 8:1632
7. Janowski JA et al (2021) Acute effects of dry needling on myofascial trigger points in the triceps surae of ballet dancers: a pilot randomized controlled trial. Int J Sports Phys Ther 16
8. Hong C-Z (2002) New trends in myofascial pain syndrome. Zhonghua Yi Xue Za Zhi (Taipei) 65:501–512

9. Ge H-Y, Fernández-de-las-Peñas C, Yue S-W (2011) Myofascial trigger points: spontaneous electrical activity and its consequences for pain induction and propagation. Chin Med 6:13
10. Shah JP et al (2008) Biochemicals associated with pain and inflammation are elevated in sites near to and remote from active myofascial trigger points. Arch Phys Med Rehabil 89:16–23
11. Chen JT et al (2001) Inhibitory effect of dry needling on the spontaneous electrical activity recorded from myofascial trigger spots of rabbit skeletal muscle. Am J Phys Med Rehabil 80:729–735
12. Cagnie B et al (2012) The influence of dry needling of the trapezius muscle on muscle blood flow and oxygenation. J Manip Physiol Ther 35:685–691
13. Skorupska E, Rychlik M, Samborski W (2015) Intensive vasodilatation in the sciatic pain area after dry needling. BMC Complement Altern Med 15:72
14. Choo HC et al (2017) Reliability of laser Doppler, near-infrared spectroscopy and Doppler ultrasound for peripheral blood flow measurements during and after exercise in the heat. J Sports Sci 35:1715–1723
15. Barstow TJ (2019) Understanding near infrared spectroscopy and its application to skeletal muscle research. J Appl Physiol 126:1360–1376
16. Moreira DG et al (2017) Thermographic imaging in sports and exercise medicine: a delphi study and consensus statement on the measurement of human skin temperature. J Therm Biol 69:155–162
17. Gliemann L, Mortensen SP, Hellsten Y (2018) Methods for the determination of skeletal muscle blood flow: development, strengths and limitations. Eur J Appl Physiol 118:1081–1094
18. Jiménez-Sánchez C et al (2021) Effects of dry needling on biomechanical properties of the myofascial trigger points measured by myotonometry: a randomized controlled trial. J Manip Physiol Ther 44:467–474
19. Valera-Calero JA et al (2023) Changes in stiffness at active myofascial trigger points of the upper trapezius after dry needling in patients with chronic neck pain: a randomized controlled trial. Acupunct Med 41:121–129
20. Lettner J et al (2024) Evaluating the reliability of MyotonPro in assessing muscle properties: a systematic review of diagnostic test accuracy. Medicina (B Aires) 60:851
21. Aird L, Samuel D, Stokes M (2012) Quadriceps muscle tone, elasticity and stiffness in older males: reliability and symmetry using the MyotonPRO. Arch Gerontol Geriatr 55:e31–e39
22. Chino K, Akagi R, Dohi M, Fukashiro S, Takahashi H (2012) Reliability and validity of quantifying absolute muscle hardness using ultrasound elastography. PLoS One 7:e45764
23. Melo ASC, Cruz EB, Vilas-Boas JP, Sousa ASP (2022) Scapular dynamic muscular stiffness assessed through myotonometry: a narrative review. Sensors 22:2565
24. Sánchez-Infante J, Bravo-Sánchez A, Jiménez F, Abián-Vicén J (2021) Effects of dry needling on mechanical and contractile properties of the upper trapezius with latent myofascial trigger points: a randomized controlled trial. Musculoskelet Sci Pract 56:102456
25. Kearns GA et al (2023) Lack of standardization in dry needling dosage and adverse event documentation limits outcome and safety reports: a scoping review of randomized clinical trials. J Manual Manipul Therapy 31:72–83
26. Sánchez-Infante J, Navarro-Santana MJ, Bravo-Sánchez A, Jiménez-Diaz F, Abián-Vicén J (2021) Is dry needling applied by physical therapists effective for pain in musculoskeletal conditions? A systematic review and meta-analysis. Phys Ther 101
27. Gattie E, Cleland JA, Snodgrass S (2017) The effectiveness of trigger point dry needling for musculoskeletal conditions by physical therapists: a systematic review and meta-analysis. J Orthop Sports Phys Ther 47:133–149
28. Mejuto-Vázquez MJ, Salom-Moreno J, Ortega-Santiago R, Truyols-Domínguez S, Fernández-de-las-Peñas C (2014) Short-term changes in neck pain, widespread pressure pain sensitivity, and cervical range of motion after the application of trigger point dry needling in patients with acute mechanical neck pain: a randomized clinical trial. J Orthop Sports Phys Ther 44:252–260
29. Fernández-de-Las-Peñas C, Nijs J (2019) Trigger point dry needling for the treatment of myofascial pain syndrome: current perspectives within a pain neuroscience paradigm. J Pain Res 12:1899–1911

30. León-Hernández JV et al (2016) Immediate and short-term effects of the combination of dry needling and percutaneous TENS on post-needling soreness in patients with chronic myofascial neck pain. Braz J Phys Ther 20:422–431
31. Koppenhaver SL et al (2017) The association between dry needling-induced twitch response and change in pain and muscle function in patients with low back pain: a quasi-experimental study. Physiotherapy 103:131–137
32. Perreault T, Dunning J, Butts R (2017) The local twitch response during trigger point dry needling: is it necessary for successful outcomes? J Bodyw Mov Ther 21:940–947
33. Fernández-de-las-Peñas C et al (2022) The importance of the local twitch response during needling interventions in spinal pain associated with myofascial trigger points: a systematic review and meta-analysis. Acupunct Med 40:299–311
34. Walton D et al (2011) Reliability, standard error, and minimum detectable change of clinical pressure pain threshold testing in people with and without acute neck pain. J Orthop Sports Phys Ther 41:644–650
35. de Oliveira AK, Dibai-Filho AV, Soleira G, Machado ACF, de Jesus Guirro RR (2021) Reliability of pressure pain threshold on myofascial trigger points in the trapezius muscle of women with chronic neck pain. Rev Assoc Med Bras 67:708–712
36. Chiarotto A et al (2019) Measurement properties of visual analogue scale, numeric rating scale, and pain severity subscale of the brief pain inventory in patients with low back pain: a systematic review. J Pain 20:245–263
37. Sánchez-Infante J, Bravo-Sánchez A, Jiménez F, Abián-Vicén J (2021) Effects of dry needling on muscle stiffness in latent myofascial trigger points: a randomized controlled trial. J Pain 22:817–825
38. Kużdżał A et al (2025) Dry needling in sports and sport recovery: a systematic review with an evidence gap map. Sports Med 55:811–844
39. Hadizadeh M, Rahimi A, Javaherian M, Velayati M, Dommerholt J (2021) The efficacy of intramuscular electrical stimulation in the management of patients with myofascial pain syndrome: a systematic review. Chiropr Man Therap 29:40
40. Baumann AN, Fiorentino A, Oleson CJ, Leland JM (2023) The impact of dry needling with electrical stimulation on pain and disability in patients with musculoskeletal shoulder pain: a systematic review and meta-analysis of randomized controlled trials. Cureus. https://doi.org/10.7759/cureus.41404
41. Garcia-de-Miguel S et al (2020) Short-term effects of PENS versus dry needling in subjects with unilateral mechanical neck pain and active myofascial trigger points in levator scapulae muscle: a randomized controlled trial. J Clin Med 9:1665
42. Hernandez JVL, Calvo-Lobo C, Zugasti AM-P, Fernandez-Carnero J, Beltran Alacreu H (2021) Effectiveness of dry needling with percutaneous electrical nerve stimulation of high frequency versus low frequency in patients with myofascial neck pain. Pain Physician 24:135–143
43. Brady S, McEvoy J, Dommerholt J, Doody C (2014) Adverse events following trigger point dry needling: a prospective survey of chartered physiotherapists. J Manual Manipul Therapy 22:134–140
44. Boyce D et al (2020) Adverse events associated with therapeutic dry needling. Int J Sports Phys Ther 15:103–113
45. Halle JS, Halle RJ (2016) Pertinent dry needling considerations for minimizing adverse effects—part one. Int J Sports Phys Ther 11:651–662
46. Trybulski R et al (2026) Physiological correlates and predictors of vasovagal responses following dry needling in myofascial pain syndrome: a controlled observational study. Ther Adv Musculoskelet Dis 18

Chapter 7
Multimodal Protocols

Abstract This chapter examines multimodal recovery protocols as integrated biological strategies rather than simple combinations of techniques. It focuses on how thermal and mechanical stimuli can be sequenced or combined to influence microvascular reactivity, interstitial fluid movement, nociception, tissue compliance, and recovery kinetics. Particular attention is given to cold-compression, heat-compression, and contrast-based approaches, as well as to oscillatory vascular response models that help explain why temporal patterns of constriction, dilation, reperfusion, and drainage may be as important as mean blood flow. The chapter also outlines the measurement architecture required to evaluate these protocols rigorously, including perfusion, vascular reactivity, temperature, oxygenation, neuromechanical properties, and symptom outcomes. Practical guidance is provided on protocol design, standardization, technical implementation, and interpretation, allowing readers to understand when multimodal interventions are physiologically justified and how they can be applied with greater precision in athletic contexts.

7.1 Why Multimodal Protocols Deserve a Separate Framework

Multimodal recovery should be defined as the deliberate co-application or sequencing of two or more physical stimuli to shape microvascular reactivity, interstitial fluid transport, nociception, and tissue mechanical state more precisely than a single modality can achieve [1]. That framing is necessary because systematic reviews show that common recovery modalities rarely dominate across all outcomes, which means protocol value is usually endpoint-specific rather than universal [2]. Therefore, we must treat multimodal protocols as biological timing systems that attempt to control when vessels constrict, when they dilate, when fluid is displaced, and when tissue compliance is restored [3].

Single modalities usually bias the system in one dominant direction, because cold favors vasoconstriction and analgesia, heat favors vasodilation and tissue extensibility, and compression primarily modifies venous outflow, interstitial pressure, and

R. Trybulski, *Biomarker-Guided Physical Recovery Interventions*,
SpringerBriefs in Forensic and Medical Bioinformatics,
https://doi.org/10.1007/978-981-92-0631-5_7

lymphatic evacuation [4, 5]. Combining modalities becomes rational when the clinician wants to uncouple outcomes that do not always travel together, such as reducing pain without prolonging local congestion, or increasing perfusion without losing edema control [6]. This also explains why multimodal protocols are better judged by a panel of endpoints than by one variable, because soreness, force, perfusion, stiffness, oxygenation, and swelling often recover at different rates [2, 7].

Practical Box

In athlete populations, cold-water immersion, contrast water therapy, compression, and massage all show some recovery utility, but the strongest effects are often on soreness and perceived fatigue rather than on every performance variable [2, 7]. *This pattern is important for combat-sport and tournament settings, where the immediate target may be symptom control or limb readiness rather than full biological repair* [8].

Highlights

- *Multimodal protocols should be selected to match a dominant biological problem, not because more treatment automatically means better treatment.*
- *The central design question is not which modality is best, but which combination best fits the desired vascular, fluid, neural, and mechanical response.*

7.2 Thermal-Mechanical Coupling

7.2.1 Cold Plus Compression

Cold-compression protocols are mechanistically attractive because cooling lowers tissue temperature and pain sensitivity while compression augments external pressure, modifies transmural gradients, and can improve fluid evacuation [4, 9]. At the vascular level, local cooling decreases nitric-oxide-supported vasodilator tone, enhances adrenergic vasoconstrictor influence, and can reduce skin blood flow to very low levels, whereas compression adds a mechanical component that may accelerate venous and lymphatic clearance despite reduced arterial inflow [5]. The practical implication is that cold plus compression is most logical when the dominant goals are containment of pain, heaviness, or swelling in the early post-load period, rather than maximization of nutritive perfusion during the intervention itself [6, 10].

The evidence base for combined cold and compression is clinically stronger than for many other multimodal pairings, although much of the literature historically comes from postoperative or injury settings rather than elite sport [10, 11]. More recent athlete data are encouraging, because cryocompression after plyometric loading has been reported to reduce soreness, heaviness, some inflammation markers, and low-frequency force depression, while improving selected recovery outcomes compared with passive recovery [12]. A randomized crossover trial in amateur football players also found that cold-compression was superior to ice or placebo for later perfusion, muscle stiffness, biochemical damage markers, reactive strength index, and perceived recovery after plyometric exercise [13].

7.2.2 Heat Plus Compression

Heat-compression protocols serve a different purpose, because heat increases blood flow, metabolism, and connective-tissue extensibility, while compression can shape where that additional fluid movement goes and may prevent simple dependent pooling [3, 4]. Local warming produces a biphasic vasodilator response in skin, with an early sensory-nerve-mediated peak followed by a plateau that depends largely on nitric oxide signaling, so heat can be used not only as a treatment but also as a physiological probe of microvascular reserve [3, 14]. Because high local temperatures also attenuate vasoconstrictor responsiveness, heat-compression is more rational in stiffness-dominant or low-perfusion states than in an acutely swollen tissue that still requires early containment [15].

Athlete studies using device-based heat-compression indicate that relatively short applications can change pain threshold, perfusion, force output, and myotonometric properties, which supports its role as a targeted priming tool before later movement or manual work [8, 16]. The most defensible biological interpretation is not that heat "heals" tissue in a generic sense, but that it transiently increases vascular conductance and tissue compliance, thereby changing the terrain in which later recovery processes occur [4].

7.2.3 Contrast Heat-Cold Plus Pressure and Drainage

Contrast protocols can be understood as attempts to impose alternating vasoconstrictor and vasodilator phases rather than holding the tissue in a single vascular state [8, 17]. This makes them conceptually attractive for tissues that are neither purely congested nor purely underperfused, because the desired output may be repeated filling and emptying cycles rather than maximal average perfusion [18, 19]. However, systematic review data show that contrast water therapy is mainly superior to passive rest and is not consistently superior to other active recovery interventions, which means its advantage is plausibly contextual rather than absolute [7, 17].

A useful design extension is to pair contrast thermal cycles with mechanical pressure or later drainage-oriented techniques, because thermal alternation influences arteriolar tone while pressure and drainage influence venous-lymphatic return [20–22]. This is supported by athlete studies in which contrast pressure protocols modified perfusion and muscle biomechanical variables, and by separate data showing that manual lymphatic drainage and intensive massage affect postexercise recovery through partly different mechanisms [22, 23]. The practical interpretation is that multimodal sequencing should distinguish between an arteriolar objective and a drainage objective, because these are related but not identical targets [1, 20].

Recent randomized controlled evidence in combat sport athletes further supports this rationale. In a single-blind trial comparing compression contrast therapy and dry needling after isometric fatigue of forearm muscles, contrast compression significantly improved perfusion and muscle tension in the early post-intervention window, with effects persisting up to 24 h [24]. The combined protocol demonstrated superior short-term perfusion enhancement compared to isolated modalities, highlighting the importance of mechanical–thermal coupling in microvascular recovery kinetics [24].

It is equally important to note that contrast water immersion alone, without added external compression, has demonstrated meaningful recovery value. Recent randomized evidence in combat sport athletes showed that alternating hot and cold water immersion significantly reduced muscle stiffness and elasticity while improving pressure pain threshold and isometric strength within 5 min and up to 1 h post-intervention [25]. Importantly, these effects were comparable to compression-based contrast protocols in most biomechanical outcomes, suggesting that thermal oscillation itself constitutes a potent physiological stimulus independent of mechanical pressure [25].

Practical Box

In combat-sport athletes, acute cold, heat, and contrast pressure interventions have been studied on forearm muscles with outcomes including perfusion, elasticity, tension, and strength, which makes this line of work directly relevant to your book's biomarker-guided positioning [8, 16]. *In later recovery windows, manual lymphatic drainage and intense sports massage appear to diverge biologically, with drainage showing stronger effects on pain and inflammatory markers and intense massage showing stronger effects on perfusion* [22].

Highlights

- *Cold-compression is containment-oriented, heat-compression is reperfusion/compliance-oriented, and contrast-pressure protocols are oscillation-oriented.*

- *The design decision is whether the tissue currently needs less inflow, more inflow, or better cycling between the two.*

7.3 Oscillatory Vascular Response Models

7.3.1 Vasomotion, Flowmotion, and Externally Imposed Oscillation

The vascular rationale for multimodal protocols becomes stronger when interpreted through flowmotion, because microvascular perfusion is not steady but oscillatory, with frequency components linked to endothelial, neurogenic, myogenic, respiratory, and cardiac influences [18, 26]. In skin blood flow analysis, the commonly discussed low-frequency bands are approximately 0.005–0.02 Hz for endothelial activity, 0.02–0.05 Hz for neurogenic activity, and 0.05–0.15 Hz for myogenic activity, which gives a practical interpretive scaffold for multimodal interventions [18, 26]. Accordingly, alternating thermal and mechanical stimuli can be conceptualized as attempts to entrain or amplify useful vascular oscillations rather than merely to raise or lower average perfusion [19, 27].

This model is especially relevant because endothelial and neurogenic oscillation amplitudes increase during certain reactivity tests, including post-occlusive reactive hyperemia and pressure-related responses, which means the response pattern contains mechanistic information that mean perfusion alone cannot provide [19, 28]. This is an important conceptual upgrade, because it turns multimodal recovery from a stacked intervention model into a signal-processing model in which timing, repetition, and rebound matter [29].

7.3.2 Reactive Hyperemia, Rebound Perfusion, and Timing Windows

Reactive hyperemia provides a useful biological analogue for recovery design, because transient restriction followed by release produces a rebound perfusion response whose slope, peak, and oscillatory tail reflect microvascular competence [30, 31]. Local heating provides a second useful analogue, because it reveals a sensory-nerve-mediated early phase and a nitric-oxide-dependent plateau, thereby separating rapid neurovascular responsiveness from sustained endothelial reserve [3, 14]. A multimodal protocol that alternates cold, heat, and pressure can therefore be interpreted as repeatedly challenging those same response systems in a treatment context rather than a pure diagnostic context [8].

Importantly, recent crossover research examining contrast compression therapy in mixed martial arts athletes demonstrated that hyperemic responses were strictly ipsilateral. While significant increases in perfusion were observed on the treated limb, no contralateral microvascular response was detected. This finding suggests that the rebound perfusion effect of contrast-compression therapy is predominantly local rather than centrally mediated, reinforcing the importance of regional vascular targeting in protocol design [32].

The key point is that the biologically meaningful variable may be the pattern of reperfusion and oscillation rather than the largest instantaneous increase in flow [19, 29]. This is one reason why a protocol can improve soreness or heaviness without producing dramatic gains in jump performance, because symptom relief, fluid redistribution, and true contractile recovery are partially separable processes [2, 12].

Practical Box

The oscillatory model helps explain why contrast methods can outperform passive rest yet still fail to outperform every other active recovery method, because the gain may occur in vascular timing or symptom kinetics rather than in all performance outputs [17]. *It also explains why future sports-recovery studies should report wavelet or spectral descriptors whenever possible instead of only pre-post perfusion means* [29, 33].

Highlights

- *Oscillation is not noise in the microcirculation; it is part of the physiology being regulated.*
- *Multimodal protocols can be interpreted as external attempts to shape vascular oscillation, rebound, and drainage timing.*

7.4 Measurement Architecture for Multimodal Protocols

7.4.1 Perfusion and Vascular Reactivity

Perfusion is not a single measure but a family of related measures, and we should distinguish resting flux, vascular reactivity, and oscillatory composition [34]. Laser Doppler flowmetry and laser speckle contrast imaging are most useful when interpreted as standardized relative measures of cutaneous perfusion and reactivity rather than as direct absolute measures of intramuscular nutritive flow [34]. This distinction matters because cutaneous and intramuscular beds may respond differently, so a robust multimodal protocol should pair skin perfusion data with either deeper imaging, oxygenation data, or functional outcomes [35, 36].

For vascular reactivity, post-occlusive reactive hyperemia and local thermal hyperemia are particularly useful because both are considered reliable noninvasive tests when methodology is standardized [30]. Cutaneous vascular conductance, calculated as flux divided by mean arterial pressure and often normalized to maximal vasodilation, is preferable to raw flux when between-session comparison is important [37]. For oscillatory analysis, forearm laser Doppler recordings should generally last at least 10 min for myogenic analysis and 10–15 min for neurogenic and endothelial low-frequency bands, otherwise the physiological spectrum is under-sampled [33].

7.4.2 Temperature and Oxygenation

Skin temperature should be measured because thermal dose is a biological input, not a decorative variable, and infrared thermography is a useful non-contact tool when room conditions, acclimation, and region-of-interest definitions are tightly controlled [38]. Thus, we argue that we should verifying whether the intended thermal exposure actually occurred and how long its surface signature persists [39, 40]. Near-infrared spectroscopy (NIRS) adds a complementary layer because it tracks muscle oxygenation and reoxygenation semi-quantitatively, thereby helping distinguish a tissue that is merely warm from one that is actually reoxygenating more effectively [36].

NIRS is especially useful when the clinician wants to track recovery kinetics rather than just an end-state value, because reoxygenation speed after loading may be more informative than a resting saturation number [41]. Portable muscle oxygenation monitors show moderate-to-good reliability under many exercise conditions, but reliability worsens at higher intensities and movement artifact remains a practical limitation [42].

7.4.3 Neuromechanical and Symptom Endpoints

Neuromechanical endpoints are indispensable in multimodal work because a protocol can change perfusion without improving tissue mechanical behavior, and the opposite can also occur [8]. Myotonometric indices such as tone, stiffness, and elasticity are attractive because systematic review evidence supports myotonometry as a valid and reliable complementary tool for viscoelastic muscle assessment, although validation remains population- and site-dependent [43]. Pressure pain threshold, soreness scales, perceived recovery, and strength tests should be collected alongside mechanical measures, because symptom relief and contractile readiness are clinically meaningful even when perfusion data look favorable [2].

Massage and drainage studies reinforce this multidomain requirement, because local blood flow, skin temperature, pain, biochemical markers, and strength do not move in lockstep after intervention [22, 44]. Accordingly, the minimum defensible endpoint package for multimodal protocols is one vascular measure, one thermal or oxygenation measure, one neuromechanical measure, and one functional or symptom measure.

Tips

- *For low-frequency flowmotion analysis with laser Doppler flowmetry, forearm recordings should usually be at least 10 min for myogenic bands and 10–15 min for neurogenic/endothelial bands, under tightly standardized environmental conditions* [33].
- *For local thermal hyperemia testing, 42 °C is a common physiological testing temperature, and 43 °C with sodium nitroprusside is commonly used to obtain maximal vasodilation for normalization* [37].
- *Infrared thermography should be interpreted only after controlling acclimation, room conditions, and recent activity, because methodological heterogeneity materially alters skin temperature maps* [38].
- *NIRS is most defensible for within-athlete trends in muscle oxygenation and reoxygenation kinetics rather than as a stand-alone absolute proxy for nutritive flow* [36].

Typical Values

- *Common low-frequency LDF bands are approximately 0.005–0.02 Hz for endothelial activity, 0.02–0.05 Hz for neurogenic activity, and 0.05–0.15 Hz for myogenic activity, with respiratory and cardiac components at higher frequencies* [18, 26].
- *In local heating studies, healthy skin can approach roughly 80–90% of maximal cutaneous vascular conductance at 42 °C, but the nitric-oxide-dependent plateau varies substantially by phenotype and pathology, so internal normalization is preferable to cross-study comparison* [15, 45, 46].
- *Research examples of pressure-time dosing include 25 mmHg for 30 min in ankle lymph-flow experiments, 40–60 mmHg for 2 h/day in sequential pneumatic compression for lymphedema, and device-specific sports protocols ranging from 3–6 min at 3 °C and 75 mmHg to 10–20 min of heat/cold compression* [16, 20, 47].

- *These values are research examples from specific tissues, devices, and clinical aims, so they should be treated as starting anchors rather than universal prescriptions.*

7.5 Practical Protocol Archetypes

A concise practical framework is to classify multimodal protocols by dominant objective rather than by device name. When pain, heaviness, swelling, or late-session congestion dominate, a cold-compression archetype is most coherent [10]. When low perfusion, residual stiffness, or pre-activity tissue priming dominate, a heat-compression archetype is more coherent [4]. When the goal is to restore readiness across repeated exposures, especially where neither pure containment nor pure vasodilation is ideal, a contrast-pressure archetype becomes plausible [17].

A reasonable inference from the physiology is that clinicians can also sequence modalities across time, using early cold-compression when containment is needed and later drainage-oriented or heat-assisted work when fluid mobilization and compliance become the priority [20, 22]. This time-sequenced approach is more biologically defensible than repeatedly applying the same modality regardless of recovery phase [6, 9].

Importantly, ischemic preconditioning (IPC) should be distinguished from traditional blood flow restriction (BFR) training, as IPC involves brief suprasystolic occlusion–reperfusion cycles applied without concurrent exercise and may serve not only as a conditioning stimulus but also as a recovery-oriented intervention, with emerging evidence indicating benefits for post-exercise perfusion, soreness attenuation, and functional restoration [48].

7.6 Standardization, Safety, and Responder Phenotypes

The strongest threat to multimodal research is poor standardization, because temperature, pressure, exposure time, body position, limb elevation, timing after exercise, and outcome timing can all materially change the biological response [4]. Responder heterogeneity should be expected rather than treated as noise, because baseline endothelial reserve, sympathetic tone, adiposity, skin characteristics, tissue depth, and the current phase of recovery all influence observed responses [1]. For that reason, we should explicitly recommend within-subject designs, repeated measures, and standardized normalization procedures whenever possible.

Safety should also be framed biologically rather than generically, because tissue that is acutely numb, ischemic, highly edematous, recently injured, or poorly perfused may not tolerate the same thermal-pressure load as healthy postexercise muscle [4].

Therefore, multimodal protocols are most defensible when they are phenotype-led, dose-specified, and biomarker-verified, not when they are marketed as universally superior recovery routines.

Practical Box

In real athlete contexts, the most informative multimodal endpoints are often later time points such as 24–72 h, because immediate post-treatment gains do not necessarily predict delayed stiffness, soreness, biochemical damage, or readiness [13, 22]. *Studies in combat athletes are especially valuable here because they combine local tissue loading, repeated exposure, and high practical demand for rapid readiness restoration* [8, 21].

Highlights

- *Multimodal protocols should be written as dose-defined biological algorithms, not as informal recovery routines.*
- *The winning protocol is the one that matches the dominant recovery phenotype and verifies its effect with the smallest defensible endpoint set.*

References

1. Roustit M, Cracowski J-L (2013) Assessment of endothelial and neurovascular function in human skin microcirculation. Trends Pharmacol Sci 34:373–384
2. Dupuy O, Douzi W, Theurot D, Bosquet L, Dugué B (2018) An evidence-based approach for choosing post-exercise recovery techniques to reduce markers of muscle damage, soreness, fatigue, and inflammation: a systematic review with meta-analysis. Front Physiol 9
3. Johnson JM, Kellogg DL (2010) Local thermal control of the human cutaneous circulation. J Appl Physiol 109:1229–1238
4. Malanga GA, Yan N, Stark J (2015) Mechanisms and efficacy of heat and cold therapies for musculoskeletal injury. Postgrad Med 127:57–65
5. Johnson JM, Kellogg DL (2018) Skin vasoconstriction as a heat conservation thermoeffector. Handb Clin Neurol 156:175–192
6. Kwiecien SY, McHugh MP (2021) The cold truth: the role of cryotherapy in the treatment of injury and recovery from exercise. Eur J Appl Physiol 121:2125–2142
7. Higgins TR, Greene DA, Baker MK (2017) Effects of cold water immersion and contrast water therapy for recovery from team sport: a systematic review and meta-analysis. J Strength Cond Res 31:1443–1460
8. Trybulski R et al (2024) Acute effects of cold, heat and contrast pressure therapy on forearm muscles regeneration in combat sports athletes: a randomized clinical trial. Sci Rep 14:22410

9. White GE, Wells GD (2013) Cold-water immersion and other forms of cryotherapy: physiological changes potentially affecting recovery from high-intensity exercise. Extrem Physiol Med 2:26
10. Block J (2010) Cold and compression in the management of musculoskeletal injuries and orthopedic operative procedures: a narrative review. Open Access J Sports Med 105. https://doi.org/10.2147/OAJSM.S11102
11. Waterman B et al (2012) The efficacy of combined cryotherapy and compression compared with cryotherapy alone following anterior cruciate ligament reconstruction. J Knee Surgery 25:155–160
12. Millour G, Lepers R, Coste A, Hausswirth C (2025) Effects of combining cold exposure and compression on muscle recovery: a randomized crossover study. Front Physiol 16
13. Trybulski R, Muracki J, Aldhahi MI, Kurtoğlu A, Halski T (2025) Effect of cold compression and ice therapy on muscle recovery after plyometric exercise: a randomized crossover trial. J Bodyw Mov Ther 45:1010–1019
14. Wong BJ, Fieger SM (2010) Transient receptor potential vanilloid type-1 (TRPV-1) channels contribute to cutaneous thermal hyperaemia in humans. J Physiol 588:4317–4326
15. Davison JL, Short DS, Wilson TE (2004) Effect of local heating and vasodilation on the cutaneous venoarteriolar response. Clin Auton Res 14:385–390
16. Trybulski R et al (2024) Optimal duration of cold and heat compression for forearm muscle biomechanics in mixed martial arts athletes: a comparative study. Med Sci Monit 30
17. Bieuzen F, Bleakley CM, Costello JT (2013) Contrast water therapy and exercise induced muscle damage: a systematic review and meta-analysis. PLoS One 8:e62356
18. Kvandal P et al (2006) Low-frequency oscillations of the laser Doppler perfusion signal in human skin. Microvasc Res 72:120–127
19. Tang Y, Xu F, Lei P, Li G, Tan Z (2023) Spectral analysis of laser speckle contrast imaging and infrared thermography to assess skin microvascular reactive hyperemia. Skin Res Technol 29
20. Meeusan R et al (1998) The influence of cold and compression on lymph flow at the ankle. Clin J Sport Med 8:266–271
21. Zebrowska A, Trybulski R, Roczniok R, Marcol W (2019) Effect of physical methods of lymphatic drainage on postexercise recovery of mixed martial arts athletes. Clin J Sport Med 29:49–56
22. Trybulski R et al (2025) Effects of lymphatic drainage and intense sports massage on muscle properties, damage, and function after eccentric plyometric training: a randomized controlled parallel trial. Eur J Appl Physiol. https://doi.org/10.1007/s00421-025-06101-9
23. Trybulski R et al (2024) The effects of combined contrast heat cold pressure therapy on post-exercise muscle recovery in MMA fighters: a randomized controlled trial. J Hum Kinet 94:127–146
24. Kużdżał A, Clemente FM, Kawczyński A, Ryszkiel I, Trybulski R (2024) Comparing the effects of compression contrast therapy and dry needling on muscle functionality, pressure pain threshold, and perfusion after isometric fatigue in forearm muscles of combat sports athletes: a single-blind randomized controlled trial. J Sports Sci Med 548–558. https://doi.org/10.52082/jssm.2024.548
25. Trybulski R et al (2025) Influence of contrast compression therapy and water immersion contrast therapy on biomechanical parameters of the forearm muscles in martial arts athletes. Front Physiol 16
26. Jan Y, Brienza DM, Geyer MJ (2005) Analysis of week-to-week variability in skin blood flow measurements using wavelet transforms. Clin Physiol Funct Imaging 25:253–262
27. Lindqvist A (1990) Noninvasive methods to study autonomic nervous control of circulation. Acta Physiol Scand Suppl 588:1–107
28. Humeau A, Koïtka A, Abraham P, Saumet J-L, L'Huillier J-P (2004) Spectral components of laser Doppler flowmetry signals recorded in healthy and type 1 diabetic subjects at rest and during a local and progressive cutaneous pressure application: scalogram analyses. Phys Med Biol 49:3957–3970

29. Rossi M, Carpi A, Galetta F, Franzoni F, Santoro G (2006) The investigation of skin blood flowmotion: a new approach to study the microcirculatory impairment in vascular diseases? Biomed Pharmacother 60:437–442
30. Roustit M, Cracowski J (2012) Non-invasive assessment of skin microvascular function in humans: an insight into methods. Microcirculation 19:47–64
31. Schubert V, Fagrell B (1991) Evaluation of the dynamic cutaneous post-ischaemic hyperaemia and thermal response in elderly subjects and in an area at risk for pressure sores. Clin Physiol 11:169–182
32. Trybulski R et al (2024) Assessment of ipsilateral and contralateral perfusion after contrast compression therapy of upper limb muscles in MMA athletes—a cross-over study. Front Physiol 15
33. Reynès C, Vinet A, Maltinti O, Knapp Y (2020) Minimizing the duration of laser Doppler flowmetry recordings while maintaining wavelet analysis quality: a methodological study. Microvasc Res 131:104034
34. Cracowski J, Roustit M (2016) Current methods to assess human cutaneous blood flow: an updated focus on laser-based-techniques. Microcirculation 23:337–344
35. Clijsen R et al (2020) Does the application of Tecar therapy affect temperature and perfusion of skin and muscle microcirculation? A pilot feasibility study on healthy subjects. J Altern Complement Med 26:147–153
36. Boushel R, Piantadosi CA (2000) Near-infrared spectroscopy for monitoring muscle oxygenation. Acta Physiol Scand 168:615–622
37. Wong BJ, Williams SJ, Minson CT (2006) Minimal role for H_1 and H_2 histamine receptors in cutaneous thermal hyperemia to local heating in humans. J Appl Physiol 100:535–540
38. Hildebrandt C, Raschner C, Ammer K (2010) An overview of recent application of medical infrared thermography in sports medicine in Austria. Sensors 10:4700–4715
39. Hohenauer E, Deliens T, Clarys P, Clijsen R (2019) Perfusion of the skin's microcirculation after cold-water immersion (10 °C) and partial-body cryotherapy (− 135 °C). Skin Res Technol 25:677–682
40. Rojas-Valverde D et al (2021) Short-term skin temperature responses to endurance exercise: a systematic review of methods and future challenges in the use of infrared thermography. Life (Basel) 11:1286
41. Cornelis N et al (2021) The use of near infrared spectroscopy to evaluate the effect of exercise on peripheral muscle oxygenation in patients with lower extremity artery disease: a systematic review. Eur J Vasc Endovasc Surg 61:837–847
42. Crum EM, O'Connor WJ, Van Loo L, Valckx M, Stannard SR (2017) Validity and reliability of the Moxy oxygen monitor during incremental cycling exercise. Eur J Sport Sci 17:1037–1043
43. Garcia-Bernal M-I, Heredia-Rizo AM, Gonzalez-Garcia P, Cortés-Vega M-D, Casuso-Holgado MJ (2021) Validity and reliability of myotonometry for assessing muscle viscoelastic properties in patients with stroke: a systematic review and meta-analysis. Sci Rep 11:5062
44. Mori H et al (2004) Effect of massage on blood flow and muscle fatigue following isometric lumbar exercise. Med Sci Monit 10:CR173-8
45. Pyevich M, Alexander LM, Stanhewicz AE (2022) Women with a history of preeclampsia have preserved sensory nerve-mediated dilatation in the cutaneous microvasculature. Exp Physiol 107:175–182
46. DuPont JJ, Farquhar WB, Townsend RR, Edwards DG (2011) Ascorbic acid or L-arginine improves cutaneous microvascular function in chronic kidney disease. J Appl Physiol 111:1561–1567
47. Trybulski R et al (2024) Immediate effect of cryo-compression therapy on biomechanical properties and perfusion of forearm muscles in mixed martial arts fighters. J Clin Med 13:1177
48. Trybulski R et al (2025) Cold compression and ischemic preconditioning with ice therapy enhance muscle recovery and functionality post-exercise: a randomized study. Ann Rehabil Med 49:411–425

Chapter 8
Designing Clinical Protocols

Abstract This chapter explains how clinical recovery protocols should be designed from a target-driven model in which the intervention is selected according to the biological problem to be modified and the endpoint used to verify change, rather than by tradition or modality label alone. It develops a practical dose-response framework centered on the interaction among temperature, pressure, exposure time, timing of application, and mechanical load, showing why these variables determine the physiological stimulus actually received by tissue. The chapter also details how protocol standardization, contextual control, and measurement reliability condition interpretability in both athlete monitoring and clinical decision-making. Particular attention is given to selecting valid microvascular, neuromechanical, symptom, performance, and biochemical endpoints, and to distinguishing mechanistic responses from clinically meaningful outcomes. The final aim is to provide a concise template for building reproducible, mechanism-based, and context-sensitive protocols that can be implemented, tested, and refined with precision.

8.1 Define the Recovery Target and Endpoint Hierarchy

Recovery protocols should be prescribed from a target-response model in which the intervention is selected because it changes a measurable biological state rather than because it is traditionally labeled as a recovery method [1, 2]. This is necessary because post-exercise recovery includes partially dissociated vascular, metabolic, nociceptive, and neuromechanical processes that recover on different time scales [3]. A short clinical protocol therefore starts with a single biological question, such as whether the priority is to reduce soreness, restore explosive force, normalize local perfusion, decrease edema, or improve tissue mechanical behavior [4, 5]. In this framework, recovery is not defined by the modality itself, but by the interaction between temperature, pressure, exposure time, and the specific biological target selected for modification. The modality is therefore only a vehicle through which dose is delivered.

R. Trybulski, *Biomarker-Guided Physical Recovery Interventions*,
SpringerBriefs in Forensic and Medical Bioinformatics,
https://doi.org/10.1007/978-981-92-0631-5_8

Practical Box

Cold-water immersion is often a sensible choice when soreness is the primary endpoint, because meta-analytic evidence supports better recovery of soreness than many competing modalities, whereas effects on force outcomes are less uniform [4]. *Compression garments become more logical when the main decision concerns restoration of strength or jump-related function across the next 24–48 h, because benefits are stronger for those outcomes than for creatine kinase (CK) or soreness in several datasets* [6]. *Massage-oriented strategies may still fit a protocol when the goal is symptom relief, flexibility, or local tissue comfort, but they should not be assumed to restore performance directly* [7].

Highlights

- *Start with the biological problem, not with the device.*
- *Choose the primary endpoint before selecting dose variables.*
- *Do not treat soreness, force, perfusion, and blood biomarkers as interchangeable expressions of recovery.*

8.2 Build the Dose: Temperature, Pressure, Time, and Mechanical Load

8.2.1 Temperature

Thermal dose should be defined as effective tissue heating or cooling rather than as the nominal temperature of a bath, pack, or chamber [8, 9]. The achieved intramuscular temperature change depends on tissue depth, adipose thickness, exposure duration, ambient conditions, and post-treatment rewarming, so the same external temperature does not create the same biological stimulus in every athlete [8, 10]. Cold therapy generally reduces pain, blood flow, edema, and metabolic demand, whereas heat therapy generally increases blood flow, metabolism, and connective-tissue extensibility [11, 12].

For post-exercise cold-water immersion, pooled evidence indicates that temperatures around 10–15 °C for 10–15 min are among the most reproducible options for reducing soreness [4, 13]. A more recent dose-focused network meta-analysis suggests that 5–10 °C for 10–15 min may be more favorable for CK and jump-related outcomes, whereas 11–15 °C for 10–15 min may be more favorable for soreness, which means the best thermal dose depends on the chosen endpoint [14]. Direct experimental work also shows that 15 °C immersion can outperform 5 °C immersion for some recovery outcomes, so colder is not automatically better [15].

In the present series of studies, contrast compression therapy was applied within a temperature range of 3–45 °C, alternating between cold (3 °C) and heat (45 °C), which corresponds to the upper and lower limits commonly reported in contrast-based recovery protocols. Using this thermal dose, we observed significant changes in tissue perfusion, muscle tone, stiffness, elasticity, and tissue temperature, confirming that this temperature amplitude is sufficient to induce measurable microvascular and neuromechanical responses in MMA athletes [16–18]. These findings indicate that the selected temperature range represents an effective thermal stimulus capable of eliciting biologically relevant adaptations in skeletal muscle.

Heat should be considered when the main problem is persistent force loss, impaired microvascular recovery, or a desire to preserve anabolic and regenerative signaling rather than simply blunt soreness [5, 19]. Local and whole-limb heating can increase intramuscular temperature and stimulate heat-shock and angiogenic signaling, and hot-water immersion protocols that maintain core temperature near 38.5–39.0 °C for about 25 min appear capable of improving recovery while limiting excessive heat strain [20, 21]. By contrast, repeated post-resistance-exercise cold-water immersion can attenuate anabolic signaling and long-term hypertrophy, so high-dose cooling is a better fit for competition congestion than for chronic muscle-growth blocks [22, 23].

8.2.2 *Pressure*

Pressure dose is the interface pressure delivered to the tissue, not the garment brand, fabric description, or device label [24]. This distinction matters because applied pressure varies widely across garments, body regions, and postures, and many studies still report compression exposure inadequately [25, 26]. In one controlled study, higher-pressure garments improved recovery of maximal voluntary contraction and countermovement jump more than lower-pressure garments after strenuous exercise, supporting the existence of a pressure-response relationship for at least some functional outcomes [27]. Beyond vascular effects, externally applied pressure also represents a mechanical stimulus capable of altering tissue deformation patterns, interstitial fluid dynamics, and potentially mechanotransductive signaling within muscle and connective tissue. For this reason, compression dose should be considered both a hemodynamic and a mechanical load variable.

In our randomized investigations, compression magnitude proved to be a decisive determinant of recovery-related responses. In a controlled comparison of pneumatic intermittent compression at 25 mmHg versus 100 mmHg, only the 100 mmHg protocol produced sustained improvements in tissue perfusion and muscle elasticity up to 48 h post-exercise, whereas 25 mmHg resulted in markedly attenuated effects [28]. Likewise, in a comparison of cold compression (25–75 mmHg) and ischemic preconditioning with proximal occlusion at 200 mmHg, higher-pressure applications induced significantly greater improvements in perfusion, muscle stiffness, and pressure pain threshold than lower-pressure or passive conditions [29]. These findings suggest that compression levels below approximately 30–40 mmHg are unlikely to generate robust microvascular or neuromechanical adaptations, whereas higher-pressure protocols are required to elicit clinically meaningful regenerative effects.

Custom-fitted garments appear to improve recovery more consistently than standard-sized garments, probably because they deliver the intended pressure profile with greater precision [24, 30]. Compression can also improve markers of venous return and resting muscle blood flow, but those effects do not guarantee large changes in every performance or symptom endpoint [6, 25]. For this reason, pressure should be treated as a measurable dose variable that must be matched to the target endpoint and not assumed from commercial sizing alone [24].

8.2.3 Time, Timing, and Mechanical Load

Time is both a dose variable and a biological context variable [3]. Exposure duration changes the depth and persistence of thermal and mechanical effects, but the timing of application relative to exercise, competition, sleep, and the next training session also changes what outcome is desirable [31, 32]. The metabolic and blood-flow profile of recovery depends on the type and duration of the preceding exercise, so identical recovery doses should not be expected to behave similarly after eccentric strength work, repeated-sprint activity, or prolonged exercise in the heat [33].

Controlled investigations demonstrate a clear time–dose–response pattern in compression-based recovery protocols. In cold compression, microvascular and analgesic responses increase with exposure duration but plateau at approximately 10 min, while $\geq$ 15-min applications are associated with greater CIVD and post-cooling hyperaemia [34]. Earlier experimental and randomized studies using 10–20-min compression or contrast protocols similarly showed significant improvements in perfusion, stiffness, and pressure pain threshold without requiring prolonged exposure [16, 29]. Although parts of the literature traditionally report protocols lasting 30–40 min, these findings suggest that shorter, precisely dosed applications are sufficient to elicit clinically meaningful regenerative responses.

When rapid same-day performance in the heat is the priority, even a 5-min cold-water immersion can lower core temperature and preserve subsequent high-intensity cycling performance [32]. When the next priority is restoration of late-phase rate of force development after muscle damage, hot water immersion may be preferable to cold water immersion [35]. Mechanical load after the intervention also matters because the biological signal created by cooling or heating interacts with the subsequent exercise stimulus, which is one reason chronic cooling can dampen some resistance-training adaptations [23]. These findings support a non-linear dose–response model in compression-based recovery. Biological benefit increases up to a physiological threshold, after which responses plateau or shift toward counter-regulatory vascular phenomena such as CIVD and augmented reactive hyperaemia. Extending exposure beyond the effective window therefore does not proportionally increase regenerative benefit and may alter the vascular profile of recovery.

Tips

- *Adjust cooling duration to local adipose thickness whenever possible, because standardized treatment times can undercool or overcool athletes with different tissue depth.*
- *Measure actual garment or device pressure at multiple anatomical sites and, ideally, in the posture used during treatment.*
- *Document the exact interval between the end of exercise and the start of the intervention, because this interval changes the physiological context in which the dose acts.*

Typical Values

- *Cold-water immersion for soreness management is commonly supported at about 10–15 °C for 10–15 min* [13].
- *A colder range of 5–10 °C for 10–15 min may be more suitable when the target is CK or jump-related recovery rather than soreness alone* [14].
- *One effective high-pressure garment profile reported in the literature was approximately 14.8 mmHg at the thigh and 24.3 mmHg at the calf, whereas a lower profile around 8.1 mmHg at the thigh and 14.8 mmHg at the calf was less effective in the same model* [27].
- *Hot-water immersion protocols that sustain core temperature around 38.5–39.0 °C for about 25 min represent a practical benchmark when heat is used for recovery rather than for simple comfort* [36].

8.3 Standardize Delivery and Context Control

Standardization is what converts a modality into a reproducible protocol [37]. Microvascular assessments are especially sensitive to procedure, and limb position relative to heart level can materially change reactive hyperemia responses even when systemic pressure and central sympathetic control change little [37, 38]. Repeated testing and a standardized priming exercise can improve the stability of laser Doppler measurements, which illustrates how easily measurement noise can be mistaken for physiology when procedures are not tightly controlled [39, 40].

Thermal protocols also require subject-specific standardization because adipose thickness modifies the rate and magnitude of intramuscular cooling and rewarming [8, 10]. Compression studies should standardize garment fit, body posture, body region, and the pressure landmarks that are actually measured, because pressure profiles differ across garments and across the same garment worn in different contexts [25, 26]. For local stiffness or tone assessment, the same anatomical landmark, body position, and preferably the same assessor should be retained across sessions, because myotonometric reliability is highest under tightly standardized conditions [41, 42].

Practical Box

Postural control is not a minor issue in perfusion work, because reactive hyperemia differs when limbs are elevated, horizontal, or dependent [38]. *Recent athlete-focused work on quadriceps myotonometry shows that measurements remain robust across examiners, but the cleanest serial tracking is still obtained when landmarking and procedure are fixed* [41]. *In compression research, failure to report actual pressure at the thigh and calf has been a major reason why otherwise similar studies cannot be compared directly* [26].

Highlights

- *Standardize posture, site, assessor, and timing before comparing results across visits.*
- *Treat adipose thickness and garment fit as determinants of delivered dose, not as background noise.*
- *If the delivered dose is not measured or controlled, protocol interpretation becomes speculative.*

8.4 Select Endpoints with Known Validity, Reliability, and Biological Meaning

8.4.1 Microvascular Endpoints

Microvascular endpoints are most useful when the protocol is intended to influence perfusion, oxygen delivery, edema handling, or ischemia-reperfusion behavior [37]. Post-occlusive reactive hyperemia quantifies the transient reperfusion response after arterial occlusion, and laser Doppler-based peak response and area-under-the-curve variables are among the more reproducible measures for repeat monitoring [40]. Near-infrared spectroscopy adds deeper information on muscle oxygenation and recovery kinetics, and current evidence supports its reliability when site selection and device-specific limitations are respected [43].

8.4.2 Neuromechanical Endpoints

Neuromechanical endpoints become particularly valuable when the target state is abnormal tone, increased stiffness, reduced elasticity, or pain-related guarding [42, 44]. Myotonometry estimates local oscillation frequency, dynamic stiffness, and viscoelastic behavior through a brief mechanical impulse, and several studies support acceptable reliability and responsiveness when positioning and landmarking are controlled [45]. If ultrasound shear-wave elastography is used, the protocol should define probe orientation, muscle state, and examiner experience explicitly, because reliability is often good but remains strongly protocol-dependent [46].

8.4.3 Symptom, Performance, and Biochemical Endpoints

Symptom and performance endpoints remain indispensable because a biological change only becomes clinically meaningful when it modifies pain, function, or readiness [47]. Among field tests, jump height and selected countermovement-jump or isometric mid-thigh pull variables often show acceptable reliability, but many derived variables do not, so only pre-specified robust metrics should be monitored serially [48]. Creatine kinase should be treated as a contextual secondary marker rather than a stand-alone recovery endpoint because its response is strongly influenced by exercise mode, muscle mass, sex, ethnicity, climate, and large interindividual variability [49].

Blood biomarkers are therefore most useful when interpreted alongside symptoms and function rather than in isolation [1]. The key methodological rule is that every endpoint used in a protocol should have a known measurement error and a clear biological interpretation [47].

Tips

- *For post-occlusive reactive hyperaemia (PORH), prioritize peak perfusion and area-under-the-curve variables over time-to-peak when the goal is serial monitoring, because they are more reproducible* [40].
- *For near-infrared spectroscopy (NIRS), keep probe location, fixation, and limb condition identical between sessions and interpret trends in kinetics more confidently than absolute cross-device values* [43].

- *For MyotonPRO, use site-specific minimal detectable change values instead of universal cutoffs whenever serial decisions are made* [45].
- *For performance monitoring, choose a small set of robust outputs such as jump height or selected force variables and ignore unstable derived measures* [48].

Typical Values

- *A practical acceptance rule in team-sport monitoring is to prefer variables with between-day coefficient of variation at or below about 10% and intraclass correlation at or above about 0.80* [48].
- *Changes in CK should be interpreted as context-dependent evidence of muscle membrane disruption rather than as a direct readout of readiness or tissue healing* [49].
- *Microvascular and neuromechanical measures are most informative when they are paired with a time-matched symptom or performance endpoint* [1].

8.5 Build Multimodal Protocols Only When the Sequence Has a Mechanistic Rationale

Multimodal protocols should be additive by design rather than accumulative by habit [50]. When two or more modalities are combined, each component should have a specific mechanistic role and a defined measurement window, otherwise the protocol becomes impossible to optimize or falsify [23]. This principle is especially important in short protocols because complexity that is not linked to endpoint logic adds noise without adding explanatory value [1].

Recent athlete studies suggest that compression-contrast therapy, lymphatic drainage, massage, and dry needling can each modify local perfusion, pain pressure threshold, or tissue mechanics, but the magnitude and persistence of these effects differ across modalities and athlete populations [50–52]. A sensible sequence could therefore use an early analgesic or cooling component when the short-term need is soreness control and a later heat or compression component when the target shifts toward perfusion or force restoration [4, 35]. Routine inclusion of high-dose cooling inside a protocol aimed at maximizing hypertrophy or remodeling remains a mechanistic mismatch even if it improves short-term comfort [22].

Practical Box

In combat-sport settings, dry needling has been associated with short-term improvements in perfusion, pressure pain threshold, muscle power, and biomechanical variables, which makes it more suitable as a localized adjunct than as a universal background treatment [52, 53]. *Manual lymphatic drainage may fit a multimodal design when the endpoint hierarchy includes swelling or enzyme-resolution variables, whereas sports massage may fit better when the goal is symptom relief or local perfusion support* [54]. *Compression plus thermal sequencing should be justified by the next performance demand, not by the assumption that more modalities must be better* [50].

Highlights

- *Do not combine modalities unless each one has a specific job/target.*
- *Sequence modalities around the next biological target and the next performance demand.*
- *Remove components that do not improve the primary endpoint beyond measurement noise.*

8.6 A Minimal Protocol Template for Clinical and Athlete Use

A dense and clinically useful protocol can usually be built in five steps: define the biological problem, choose one primary endpoint, select the smallest plausible effective dose, standardize delivery and measurement, and adapt only after the observed change exceeds measurement noise [1]. Measurement error must be considered explicitly because a statistically significant group effect does not automatically represent a meaningful change in an individual athlete or patient [47]. Responder logic is especially important in thermal and compression interventions because delivered dose and biological response vary with tissue depth, adiposity, exercise modality, garment fit, and the subsequent training objective [10].

In practice, protocol tables should therefore state the indication, target endpoint, modality dose, application timing, standardization requirements, re-test window, and stop-or-adapt rules [26]. This type of structured reporting aligns with repeated calls for better standardization in compression, microvascular, and imaging-based recovery research [26]. Precision in recovery design therefore depends on aligning biological target, delivered dose, measurement validity, and decision criteria within the same structured framework.

Tips

- *Indication: state the clinical or sport problem in one line.*
- *Primary endpoint: choose one variable that matches the decision to be made.*
- *Dose: record temperature, pressure, duration, coverage area, and interval from exercise.*
- *Standardization: record posture, site, assessor, device settings, and any tissue-depth modifiers.*
- *Re-test rule: repeat the primary endpoint at a predefined window and change the protocol only if the response exceeds expected noise.*

Highlights

- *The shortest valid protocol is usually better than a complex protocol with unknown dose delivery.*
- *Every protocol should be written so that another clinician can reproduce it exactly.*
- *Clinical translation depends on dose precision, endpoint precision, and decision precision occurring in the same protocol.*

References

1. Lee EC et al (2017) Biomarkers in sports and exercise: tracking health, performance, and recovery in athletes. J Strength Cond Res 31:2920–2937
2. Haller N et al (2023) Blood-based biomarkers for managing workload in athletes: considerations and recommendations for evidence-based use of established biomarkers. Sports Med 53:1315–1333
3. Bangsbo J, Hellsten Y (1998) Muscle blood flow and oxygen uptake in recovery from exercise. Acta Physiol Scand 162:305–312
4. Moore E et al (2023) Effects of cold-water immersion compared with other recovery modalities on athletic performance following acute strenuous exercise in physically active participants: a systematic review, meta-analysis, and meta-regression. Sports Med 53:687–705
5. Sabapathy M et al (2021) Effect of heat pre-conditioning on recovery following exercise-induced muscle damage. Curr Res Physiol 4:155–162
6. Brown F et al (2017) Compression garments and recovery from exercise: a meta-analysis. Sports Med 47:2245–2267
7. Dakić M et al (2023) The effects of massage therapy on sport and exercise performance: a systematic review. Sports 11:110

8. Jutte LS, Merrick MA, Ingersoll CD, Edwards JE (2001) The relationship between intramuscular temperature, skin temperature, and adipose thickness during cryotherapy and rewarming. Arch Phys Med Rehabil 82:845–850
9. Jackman JS et al (2023) Effect of hot water immersion on acute physiological responses following resistance exercise. Front Physiol 14
10. Rech N, Bressel E, Louder T (2021) Predictive ability of body fat percentage and thigh anthropometrics on tissue cooling during cold-water immersion. J Athl Train 56:548–554
11. Malanga GA, Yan N, Stark J (2015) Mechanisms and efficacy of heat and cold therapies for musculoskeletal injury. Postgrad Med 127:57–65
12. Nadler SF, Weingand K, Kruse RJ (2004) The physiologic basis and clinical applications of cryotherapy and thermotherapy for the pain practitioner. Pain Physician 7:395–399
13. Machado AF et al (2016) Can water temperature and immersion time influence the effect of cold water immersion on muscle soreness? A systematic review and meta-analysis. Sports Med 46:503–514
14. Wang H, Wang L, Pan Y (2025) Impact of different doses of cold water immersion (duration and temperature variations) on recovery from acute exercise-induced muscle damage: a network meta-analysis. Front Physiol 16
15. Vieira A et al (2016) The effect of water temperature during cold-water immersion on recovery from exercise-induced muscle damage. Int J Sports Med 37:937–943
16. Trybulski R et al (2024) Optimal duration of cold and heat compression for forearm muscle biomechanics in mixed martial arts athletes: a comparative study. Med Sci Monit 30
17. Trybulski R et al (2024) The effects of combined contrast heat cold pressure therapy on post-exercise muscle recovery in MMA fighters: a randomized controlled trial. J Hum Kinet 94:127–146
18. Trybulski R et al (2024) Immediate effect of compression contrast therapy on quadriceps femoris muscles' regeneration in MMA fighters. J Clin Med 13:7292
19. Kim K, Monroe JC, Gavin TP, Roseguini BT (2020) Local heat therapy to accelerate recovery after exercise-induced muscle damage. Exerc Sport Sci Rev 48:163–169
20. Gibson OR et al (2023) Skeletal muscle angiogenic, regulatory, and heat shock protein responses to prolonged passive hyperthermia of the human lower limb. Am J Phys Regul Integr Comp Phys 324:R1–R14
21. Kim K et al (2020) Effects of repeated local heat therapy on skeletal muscle structure and function in humans. J Appl Physiol 128:483–492
22. Fyfe JJ et al (2019) Cold water immersion attenuates anabolic signaling and skeletal muscle fiber hypertrophy, but not strength gain, following whole-body resistance training. J Appl Physiol 127:1403–1418
23. Hyldahl RD, Peake JM (2020) Combining cooling or heating applications with exercise training to enhance performance and muscle adaptations. J Appl Physiol 129:353–365
24. Brown FCW, Hill JA, Pedlar CR (2022) Compression garments for recovery from muscle damage: evidence and implications of dose responses. Curr Sports Med Rep 21:45–52
25. O'Riordan SF, McGregor R, Halson SL, Bishop DJ, Broatch JR (2023) Sports compression garments improve resting markers of venous return and muscle blood flow in male basketball players. J Sport Health Sci 12:513–522
26. Weakley J et al (2022) Putting the squeeze on compression garments: current evidence and recommendations for future research: a systematic scoping review. Sports Med 52:1141–1160
27. Hill J et al (2017) The effects of compression-garment pressure on recovery after strenuous exercise. Int J Sports Physiol Perform 12:1078–1084
28. Trybulski R et al (2025) Effect of pneumatic and cold compression on muscle performance and recovery in combat sports athletes. Sci Rep 15:44993
29. Trybulski R et al (2025) Cold compression and ischemic preconditioning with ice therapy enhance muscle recovery and functionality post-exercise: a randomized study. Ann Rehabil Med 49:411–425
30. Brown F et al (2022) Custom-fitted compression garments enhance recovery from muscle damage in Rugby players. J Strength Cond Res 36:212–219

31. Almeida AC et al (2016) The effects of cold water immersion with different dosages (duration and temperature variations) on heart rate variability post-exercise recovery: a randomized controlled trial. J Sci Med Sport 19:676–681
32. Peiffer JJ, Abbiss CR, Watson G, Nosaka K, Laursen PB (2010) Effect of a 5-min cold-water immersion recovery on exercise performance in the heat. Br J Sports Med 44:461–465
33. Saltin B, Radegran G, Koskolou MD, Roach RC (1998) Skeletal muscle blood flow in humans and its regulation during exercise. Acta Physiol Scand 162:421–436
34. Trybulski R et al (2026) Time-dependent microvascular and analgesic responses to cold compression in healthy adults: identifying the optimal cooling duration for recovery. Eur J Appl Physiol. https://doi.org/10.1007/s00421-026-06141-9
35. Benoît S et al (2024) Hot but not cold water immersion mitigates the decline in rate of force development following exercise-induced muscle damage. Med Sci Sports Exerc 56:2362–2371
36. Sautillet B et al (2024) Hot water immersion: maintaining core body temperature above 38.5 °C mitigates muscle fatigue. Scand J Med Sci Sports 34
37. Roustit M, Cracowski J (2012) Non-invasive assessment of skin microvascular function in humans: an insight into methods. Microcirculation 19:47–64
38. Krishnan A, Lucassen EB, Hogeman C, Blaha C, Leuenberger UA (2011) Effects of limb posture on reactive hyperemia. Eur J Appl Physiol 111:1415–1420
39. Tran BD et al (2016) Exercise and repeated testing improves accuracy of laser Doppler assessment of microvascular function following shortened (1-minute) blood flow occlusion. Microcirculation 23:293–300
40. Barwick A, Lanting S, Chuter V (2015) Intra-tester and inter-tester reliability of post-occlusive reactive hyperaemia measurement at the hallux. Microvasc Res 99:67–71
41. Trybulski R et al (2024) Reliability of MyotonPro in measuring the biomechanical properties of the quadriceps femoris muscle in people with different levels and types of motor preparation. Front Sports Act Living 6
42. McGowen JM et al (2024) Myotonometry is capable of reliably obtaining trunk and thigh muscle stiffness measures in military cadets during standing and squatting postures. Mil Med 189:e213–e219
43. Corral-Pérez J et al (2024) Reliability of near-infrared spectroscopy in measuring muscle oxygenation during squat exercise. J Sci Med Sport 27:805–813
44. Chuang L, Wu C, Lin K (2012) Reliability, validity, and responsiveness of myotonometric measurement of muscle tone, elasticity, and stiffness in patients with stroke. Arch Phys Med Rehabil 93:532–540
45. Muckelt PE et al (2022) Protocol and reference values for minimal detectable change of MyotonPRO and ultrasound imaging measurements of muscle and subcutaneous tissue. Sci Rep 12:13654
46. Kozinc Ž, Šarabon N (2020) Shear-wave elastography for assessment of trapezius muscle stiffness: reliability and association with low-level muscle activity. PLoS One 15:e0234359
47. Ryan S et al (2019) Measurement characteristics of athlete monitoring tools in professional Australian football. Int J Sports Physiol Perform 1–7. https://doi.org/10.1123/ijspp.2019-0060
48. Aben HGJ et al (2020) The reliability of neuromuscular and perceptual measures used to profile recovery, and the time-course of such responses following academy Rugby league match-play. Sports 8:73
49. Brancaccio P, Maffulli N, Limongelli FM (2007) Creatine kinase monitoring in sport medicine. Br Med Bull 81–82:209–230
50. Kużdżał A, Clemente FM, Kawczyński A, Ryszkiel I, Trybulski R (2024) Comparing the effects of compression contrast therapy and dry needling on muscle functionality, pressure pain threshold, and perfusion after isometric fatigue in forearm muscles of combat sports athletes: a single-blind randomized controlled trial. J Sports Sci Med 548–558. https://doi.org/10.52082/jssm.2024.548
51. Zebrowska A, Trybulski R, Roczniok R, Marcol W (2019) Effect of physical methods of lymphatic drainage on postexercise recovery of mixed martial arts athletes. Clin J Sport Med 29:49–56

52. Trybulski R, Stanula A, Żebrowska A, Podleśny M, Hall B (2024) Acute effects of the dry needling session on gastrocnemius muscle biomechanical properties, and perfusion with latent trigger points—a single-blind randomized controlled trial in mixed martial arts athletes. J Sports Sci Med 136–146. https://doi.org/10.52082/jssm.2024.136
53. Trybulski R et al (2024) Biomechanical profile after dry needling in mixed martial arts. Int J Sports Med. https://doi.org/10.1055/a-2342-3679
54. Schillinger A et al (2006) Effect of manual lymph drainage on the course of serum levels of muscle enzymes after treadmill exercise. Am J Phys Med Rehabil 85:516–520

Chapter 9
Translational Applications

Abstract This chapter translates recovery science into clinically and sport-relevant decision-making by linking biological targets, measurable endpoints, and implementation strategies. It examines how translational applications should be guided by the dominant recovery problem, including pain, edema, stiffness, impaired tissue mechanics, neuromuscular fatigue, and reduced performance readiness. The chapter discusses applications in clinical rehabilitation and high-performance sport, emphasizing criterion-based progression, short-turnaround recovery, and return-to-participation or return-to-performance decisions. It also addresses responder phenotypes and the need for individualized prescriptions based on repeated within-person patterns rather than generic assumptions. Particular attention is given to device-oriented implications, including monitoring technologies and intervention systems, with discussion of what they measure, how they should be implemented, and their methodological limitations. Finally, the chapter proposes principles for standardization and endpoint selection, highlighting multidomain assessment, reporting quality, and the need to match outcomes and timing to the specific translational question being addressed.

9.1 Translational Logic: From Biological Target to Implementation

9.1.1 Why Recovery Decisions Should Be Target-Specific Rather than Modality-Driven

Recovery should be prescribed against a defined biological target, because post-exercise and post-injury recovery are multifactorial processes involving soreness, fatigue, inflammation, tissue mechanics, autonomic perturbation, and performance restoration rather than a single uniform state [1, 2]. This is why we should begin by asking whether the immediate goal is analgesia, edema control, restoration of

R. Trybulski, *Biomarker-Guided Physical Recovery Interventions*,
SpringerBriefs in Forensic and Medical Bioinformatics,
https://doi.org/10.1007/978-981-92-0631-5_9

range of motion, improvement of local tissue mechanics, acceleration of neuromuscular readiness, or support of repeated performance, because each target privileges a different endpoint and not necessarily a different best modality [1, 3].

The literature also shows that modalities do not affect all recovery domains equally, with massage and cold exposure showing stronger effects on soreness and inflammatory markers, compression showing smaller but potentially useful effects on some recovery outcomes, and performance effects often being domain- and timing-specific [2, 4]. A translational framework therefore works best when it moves in the sequence problem—biological target—measurable endpoint—intervention dose—reassessment, because this prevents the common error of applying a familiar modality without first deciding what exactly should change [1]. In practice, this means that a modality such as cold exposure may be justified when irritability, soreness, or short-turnaround recovery is the main problem, whereas a modality such as low-load blood flow restriction is better justified when mechanical loading tolerance is low but preservation or restoration of strength is the major target [5, 6].

Practical Box

In elite soccer, incomplete recovery across 72 h is common during congested schedules, which makes target-specific recovery planning more defensible than routine modality use [7, 8]. *In combat sports, recent scoping work shows that investigators increasingly combine biochemical, physiological, and physical outcomes rather than relying on a single marker of recovery* [9].

Highlights

- *Translation starts with defining the bottleneck of recovery, not with choosing a favorite modality.*
- *The same modality can be useful for one target and irrelevant for another, so endpoint selection must come before intervention selection.*
- *Recovery status is multidimensional, so no single variable should be interpreted as a complete recovery diagnosis.*

9.2 Clinical Rehabilitation Applications

9.2.1 Pain-Dominant and Edema-Dominant Presentations

In clinical rehabilitation, cold-based interventions are most defensible when the main short-term objective is to reduce pain, irritability, or swelling, because cooling can reduce pain and blood flow acutely even though evidence for accelerated tissue regeneration in humans remains limited [3, 10]. This distinction matters because the clinician should not promise faster biological healing when the more evidence-based benefit is often symptom modulation that facilitates earlier movement, tolerance to exercise, or improved participation in rehabilitation sessions [3, 11]. Postoperative literature also suggests that adding compression to cryotherapy can improve pain and swelling control in some contexts, with evidence after knee procedures showing advantages for pain, swelling, and functional progression compared with simpler approaches [12, 13].

In patients presenting with pain accompanied by edema, not only the indication but also the duration of cold-compression appears clinically relevant. In a randomized, sham-controlled trial, Trybulski et al. demonstrated a clear time-dependent response: although analgesia (PPT increase) plateaued at approximately 10 min, longer applications ($\geq$ 15–20 min) were associated with a greater incidence of cold-induced vasodilation (CIVD) and more pronounced post-intervention reactive hyperemia [14]. From a microcirculatory perspective, excessive rebound vasodilation may increase hydrostatic pressure and fluid filtration into the interstitium, potentially counteracting edema-control goals [14]. Therefore, when pain coexists with swelling, shorter cryo-compression protocols (~ 10 min) may represent a more physiologically justified window, providing meaningful symptom modulation while limiting exaggerated vascular rebound.

Heat-oriented strategies are better aligned with problems characterized by stiffness, reduced extensibility, and movement limitation, because heat can increase blood flow, metabolism, and connective-tissue extensibility and may help when the rehabilitation aim is to prepare tissue for subsequent exercise exposure rather than to quiet an acutely irritable state [3]. Dry needling also appears more defensible as a local adjunct for pain- and stiffness-dominant regions than as a universal whole-body recovery strategy, because current synthesis suggests more consistent effects on pain and muscle stiffness than on direct athletic performance outcomes [15]. From a translational viewpoint, this means that local symptom-dominant presentations should be treated with locally targeted interventions, whereas global recovery methods should be reserved for global recovery problems [1, 15].

9.2.2 *Load-Intolerant Muscle Dysfunction and Progression Toward Function*

When the main rehabilitation bottleneck is low load tolerance with persistent weakness or atrophy risk, low-load blood flow restriction has a more direct mechanistic rationale than passive recovery modalities because it can augment strength-oriented adaptation while keeping external mechanical loads relatively low [6]. This is especially relevant after surgery or painful musculoskeletal conditions, where the clinician may need to preserve or restore muscle function before heavier loading is clinically acceptable [6]. The translational point is that symptom-relieving modalities should not replace progressive loading, because the strongest long-term rehabilitation signal still comes from restoring force capacity, function, and task exposure [11].

Return-to-sport or return-to-function decisions should therefore be criterion-based and multidimensional rather than purely time-based, because current literature shows substantial variability in how return to sport is defined and continued ambiguity in the objective criteria used for progression [16]. Pain remains a common driver of rehabilitation progression in hamstring injury studies, but systematic review data also show the need for more objective and clinically practical progression criteria beyond symptom improvement alone [17]. Thus, translational rehabilitation combines symptom control, objective function, and graded task exposure rather than treating symptom relief as the end point itself.

Practical Box

After knee arthroplasty, compressive cryotherapy has been associated with faster gains in knee motion and greater swelling reduction than standard cryotherapy alone, which makes it useful when early range of motion (ROM) and edema are limiting rehabilitation [12]. *After anterior cruciate ligament (ACL) reconstruction and other knee procedures, postoperative device reviews support cryotherapy or cryocompression mainly for pain relief and support of early functional progression, not as substitutes for strengthening* [13].

Highlights

- *In rehabilitation, the first translational question is whether the main barrier is pain, swelling, stiffness, weakness, or task intolerance.*
- *Cooling is most evidence-based for symptom modulation, whereas load-restoring strategies are needed for durable functional change.*
- *Criterion-based progression is preferable to time-only progression.*

9.3 High-Performance Sport Applications

9.3.1 Congested Competition Schedules and Short-Turnaround Recovery

In high-performance sport, translational recovery is constrained by time, because athletes often need to re-express high neuromuscular output before biochemical restoration is complete [8, 18]. That asymmetry is especially important in team and combat sports, where creatine kinase (CK) or soreness may remain elevated even when parts of physical performance have already recovered, meaning that "recovered" depends on which endpoint is prioritized [18]. Accordingly, the aim of many short-turnaround interventions in elite sport is not full biological normalization but restoration of sufficient readiness for the next exposure with acceptable risk [1, 7].

Cold-water immersion has the strongest current evidence base among common sport recovery modalities for short-turnaround use, particularly for acute strenuous exercise in physically active populations, although its benefits are typically clearer for perceptual recovery and some performance contexts than as a universal solution for every outcome [5]. Compression garments can also support recovery in some contexts, with meta-analytic evidence showing small overall benefits and stronger effects in some strength-recovery windows, but the signal is not so large that compression should replace better-validated foundations such as training-load management, sleep, and nutrition [4]. Massage remains most defensible for soreness, flexibility, and athlete acceptance rather than for direct restoration of sprint, jump, or endurance performance [19].

9.3.2 Readiness Monitoring and Return-to-Train/ Return-to-Performance Decisions

Because no single gold-standard biomarker of post-match or post-session recovery exists, readiness decisions in elite sport should combine neuromuscular, perceptual, and physiological domains rather than depend on one isolated measure [20]. Countermovement-jump (CMJ) monitoring is useful because it is sensitive to acute neuromuscular status, but the exact variable selected matters, since alternative force-time variables can be more informative than jump height alone depending on the fatigue phenotype and the measurement setup [21, 22]. Heart-rate variability can add autonomic information, but interpretation must be standardized and athlete-specific because HRV is sensitive to measurement context, and portable devices introduce small but relevant methodological error when compared with electrocardiogram (ECG) [23].

External-load data can sharpen recovery planning because high-speed running exposure is associated with acute and residual fatigue markers after soccer match play, which helps explain why players with similar wellness scores may still need different recovery prescriptions after different match demands [24]. This is exactly where translational recovery should move beyond generic postmatch routines and toward individualized post-exposure responses based on the interaction among load, symptoms, and objective neuromuscular status [24].

Recent combat-sport studies, including work from Trybulski and colleagues, also support this multidomain logic by combining perfusion, muscle mechanics, soreness, and jump-derived performance when comparing cold compression, pneumatic compression, contrast systems, lymphatic drainage, or massage [25–27].

Practical Box

In male soccer, 72 h may be insufficient for full restoration of muscle-damage and wellness markers even when some performance outcomes improve, so recovery routines should be aligned with the next task rather than with a fixed clock alone [8]. *In female soccer, currently available review data also indicate distinct timelines across performance, physiological, and perceptual outcomes, reinforcing the need for context-specific monitoring* [28].

Highlights

- *Short-turnaround sport recovery is about restoring usable readiness, not necessarily complete physiological normalization.*
- *Cold water immersion has one of the strongest evidence bases for acute recovery use, but its effect size depends on timing and outcome domain.*
- *Readiness decisions should integrate load, symptoms, and neuromuscular or autonomic markers.*

Tips

- *For CMJ monitoring, keep the same warm-up, instructions, arm position, device, and variable set across sessions because sensitivity and reliability change with the metric chosen.*
- *For heart rate variability (HRV), standardize posture, timing, breathing conditions, and device type, and avoid mixing ECG-derived and portable-device data without validation.*

- *For postmatch planning, interpret subjective data alongside external load because high-speed exposure helps explain residual fatigue better than generic workload summaries alone.*

9.4 Responder Phenotypes and Individualized Prescription

9.4.1 Biological and Contextual Sources of Response Heterogeneity

A central translational challenge is that response heterogeneity is real, but it is often overstated when studies fail to distinguish true interindividual response differences from ordinary measurement error and day-to-day biological variability [29]. For this reason, the language of "non-responder" should be used cautiously, because precision-exercise literature argues that low sensitivity for one endpoint or one dose does not imply global non-responsiveness [30]. In recovery science, this means that different athletes or patients may diverge because of training status, injury state, sex, adipose tissue thickness, autonomic profile, pain sensitivity, local tissue state, or simply because the chosen endpoint is more or less sensitive to the intervention [31].

The practical implication is that phenotype should be defined around repeated patterns, not single observations, because one isolated abnormal score after training or treatment is insufficient to classify an athlete or patient as a specific responder type [29]. A sport-ready phenotype can therefore be operationalized from clustered information such as high soreness/low CMJ recovery, high sympathetic strain/slow HRV normalization, edema-dominant local response, or stiffness-dominant local response, provided those patterns repeat across exposures [20]. This n-of-1 logic is more defensible than universal prescriptions, especially in elite sport and chronic rehabilitation where within-person baselines are usually more informative than population averages [32, 33].

9.4.2 Practical Phenotyping for Recovery Prescription

A concise translational phenotyping model can classify individuals into four practical profiles: pain-dominant, edema/perfusion-dominant, stiffness/neuromechanics-dominant, and performance-readiness-dominant, because these profiles map more directly onto intervention goals than broad diagnostic labels alone [3]. Pain-dominant profiles may benefit more from cooling or dry needling when local irritability is the bottleneck, edema/perfusion-dominant profiles may justify compressive or cryo-compressive strategies, stiffness-dominant profiles may justify heat, manual therapy, or local needling, and performance-readiness-dominant profiles may justify short-turnaround cold water immersion or compression based on competition constraints [3, 5, 15]. The phenotype should then be re-tested after intervention, because translational work is strongest when it closes the loop between baseline pattern, targeted treatment, and measured response.

Practical Box

Two players with similar postmatch soreness can need different interventions if one has depressed CMJ measures and the other has normal jump output but persistent local stiffness. In rehabilitation, a patient who is pain-limited but not force-limited should not automatically receive the same progression as a patient whose pain is mild but whose strength restoration is poor.

Highlights

- *"Responder phenotype" should mean a repeatable within-person pattern, not a one-off postexercise value.*
- *Precision recovery requires repeated measures, comparator logic, and context-rich interpretation.*
- *Within-person baselines usually outperform universal norms for translational decision-making.*

9.5 Device-Oriented Implications

9.5.1 Monitoring Devices: What They Measure, How to Use Them, and What Is Reliable

Myotonometry is attractive for translational use because it provides non-invasive estimates of tissue mechanical behavior such as oscillation frequency, stiffness, and decrement, and recent systematic review data support generally high intra- and inter-rater reliability across many muscles when protocols are standardized [34, 35]. This makes MyotonPRO particularly useful when the we needs an objective bridge between local palpatory impressions and quantifiable tissue mechanics, although values remain site-, posture-, and muscle-state dependent [34]. Athlete-specific reliability data, including work by Trybulski and colleagues in quadriceps assessment across different sport populations, further support its practical use when assessors standardize site selection and technique [36].

Ultrasound shear-wave elastography can quantify muscle stiffness non-invasively, but the literature still shows major methodological inconsistency, which is why current reviews recommend reporting shear-wave velocity rather than converting to elastic modulus and emphasize the need for standardized scanning protocols [37]. Near-infrared spectroscopy can extend translational monitoring by estimating local muscle oxygenation dynamics, but reliability is stronger for some threshold applications than others, and interpretation can be modified by adipose tissue, region selection, and detection method [38]. Laser Doppler and laser speckle approaches are useful for peripheral microvascular reactivity, yet they are highly protocol-sensitive, and skin-based measurements should be interpreted as standardized surrogates of vascular behavior rather than direct whole-muscle recovery readouts [39].

9.5.2 Intervention Devices: Compression, Cryo-compression, Contrast Systems, and Needling-Related Implementation

Device-based interventions should be selected according to the physiological bottleneck they are expected to influence, because compression-oriented devices mainly target venous return, fluid movement, and soreness, whereas cryo-compression and contrast systems combine thermal and mechanical signaling with different time-course effects [40–42]. This is supported by recent sport studies showing that cryo-compression may improve pain and some recovery variables acutely, while pneumatic compression may have more sustained effects on selected tissue-mechanical outcomes depending on pressure and protocol design [25, 26]. Contrast heat-cold pressure systems are also promising where a clinician or performance practitioner wants to target both perfusion behavior and local tissue mechanics, but the evidence base remains smaller than for classic CWI and should be presented as emerging rather than definitive [5, 43].

Dry needling belongs in this section as a precision local tool rather than as a generalized device-equivalent recovery method, because the current sports literature indicates more consistent usefulness for local pain and stiffness reduction than for direct enhancement of global athletic performance [15]. Accordingly, the implementation standard for needling should emphasize region-specific indication, anatomical safety, symptom irritability, and its role as an adjunct that should facilitate movement or exercise rather than replace them [15].

Practical Box

In amateur football players after plyometric exercise, cold compression improved perfusion, reactive strength index, and stiffness-related outcomes relative to placebo in a recent crossover design [25]. *In combat-sport athletes, recent randomized studies suggest that pneumatic compression, cryo-compression, lymphatic drainage, and intense massage may influence different combinations of perfusion, soreness, inflammation, and tissue mechanics, which supports a target-based rather than brand-based approach* [26, 27].

Highlights

- *Monitoring devices are only useful when the practitioner understands what biological domain they actually sample and what their measurement error permits.*
- *Intervention devices should be matched to a defined physiological problem, not chosen because they are technologically sophisticated.*
- *Emerging device studies are valuable, but older evidence hierarchies still favor broader modalities such as cold water immersion for routine short-turnaround sport use.*

9.6 Standardization and Endpoint Selection

9.6.1 A Minimum Core Endpoint Set for Translational Studies and Practice

The literature does not support a single gold-standard recovery biomarker, so we should recommend a minimum multidomain endpoint set rather than a single preferred marker. For sport, a pragmatic minimum set is one subjective endpoint (for example soreness or perceived recovery), one neuromuscular endpoint (for example CMJ or isometric force), and one physiological or local-tissue endpoint chosen according to the intervention mechanism, such as HRV, perfusion reactivity, or tissue stiffness [21, 23, 34]. For rehabilitation, the minimum set should include one symptom endpoint, one function or participation endpoint, and one objective physical endpoint such as strength, ROM, or task performance, because return-to-sport and return-to-function decisions are not adequately captured by symptoms alone [16, 17].

Endpoint timing must also match the mechanism being tested, because immediate analgesic effects, 24–48 h tissue-mechanical effects, and 48–72 h biochemical or functional effects do not necessarily evolve on the same timeline [18, 25]. This is why translational studies should predefine which outcome is primary, which are mechanistic secondary outcomes, and what time points represent clinically or competitively meaningful windows [29]. Without that discipline, the same study can appear positive or negative depending only on which endpoint and time point the author chooses to emphasize.

9.6.2 Reporting Standards, Timing of Assessment, and Interpretation of Change

We should explicitly recommend reporting the dose of the intervention in physical units whenever possible, including pressure, temperature, duration, exposure frequency, limb coverage, and timing relative to exercise or rehabilitation loading, because these variables materially influence interpretation and reproducibility. For measurement endpoints, investigators and clinicians should report device model, body site, body position, assessor training, repeated-measure reliability, and the smallest detectable or practically meaningful change, because otherwise translational interpretation becomes fragile. This is especially important in recovery science, where small effects are common and false personalization can emerge when ordinary noise is mistaken for a biological response pattern [29].

A final translational recommendation is to align terminology with the return-to-sport continuum by separating return to participation, return to sport, and return to performance, because each stage demands a different threshold of symptom control, function, exposure tolerance, and performance readiness. That distinction prevents overclaiming success when an athlete has resumed activity but has not yet restored preinjury or prefatigue performance capacity [16]. This may be the most important standardization message. Endpoints should reflect the actual question being asked, not simply the variable that is easiest to collect [11].

Practical Box

A cold-water immersion study aimed at short-turnaround match readiness should not use 72-h CK as its only primary endpoint, because the practical question is next-session readiness rather than complete biomarker normalization. A rehabilitation study claiming faster return to sport should define whether the endpoint means return to training, return to competition, or return to preinjury performance level.

Highlights

- *There is no single universal recovery biomarker.*
- *The strongest translational designs use multidomain endpoint batteries and prespecified time windows.*
- *Reporting the physical dose of the intervention is as important as reporting the outcome.*

Tips

- *Define one primary endpoint before data collection and ensure it matches the stated practical objective.*
- *Standardize assessment timing relative to exercise, treatment, food intake, and time of day.*
- *Report the intervention dose in physical units and the measurement error of the monitoring tool.*
- *Prefer within-subject repeated baselines when the goal is individualized prescription.*

References

1. Kellmann M et al (2018) Recovery and performance in sport: consensus statement. Int J Sports Physiol Perform 13:240–245
2. Dupuy O, Douzi W, Theurot D, Bosquet L, Dugué B (2018) An evidence-based approach for choosing post-exercise recovery techniques to reduce markers of muscle damage, soreness, fatigue, and inflammation: a systematic review with meta-analysis. Front Physiol 9
3. Malanga GA, Yan N, Stark J (2015) Mechanisms and efficacy of heat and cold therapies for musculoskeletal injury. Postgrad Med 127:57–65
4. Brown F et al (2017) Compression garments and recovery from exercise: a meta-analysis. Sports Med 47:2245–2267
5. Moore E et al (2023) Effects of cold-water immersion compared with other recovery modalities on athletic performance following acute strenuous exercise in physically active participants: a systematic review, meta-analysis, and meta-regression. Sports Med 53:687–705
6. Hughes L, Paton B, Rosenblatt B, Gissane C, Patterson SD (2017) Blood flow restriction training in clinical musculoskeletal rehabilitation: a systematic review and meta-analysis. Br J Sports Med 51:1003–1011
7. Nédélec M et al (2012) Recovery in soccer: part I—post-match fatigue and time course of recovery in soccer. Sports Med 42:997–1015
8. Silva JR et al (2018) Acute and residual soccer match-related fatigue: a systematic review and meta-analysis. Sports Med 48:539–583
9. Trybulski R et al (2026) Postexercise physical recovery methods for combat sports: a scoping review. Int J Sports Med 47:3–23

10. Racinais S et al (2024) Cryotherapy for treating soft tissue injuries in sport medicine: a critical review. Br J Sports Med 58:1215–1223
11. Ardern CL et al (2016) 2016 consensus statement on return to sport from the first world congress in sports physical therapy. Bern Br J Sports Med 50:853–864
12. Quesnot A et al (2024) Randomized controlled trial of compressive cryotherapy versus standard cryotherapy after total knee arthroplasty: pain, swelling, range of motion and functional recovery. BMC Musculoskelet Disord 25:182
13. Martimbianco ALC et al (2014) Effectiveness and safety of cryotherapy after arthroscopic anterior cruciate ligament reconstruction. A systematic review of the literature. Phys Ther Sport 15:261–268
14. Trybulski R et al (2026) Time-dependent microvascular and analgesic responses to cold compression in healthy adults: identifying the optimal cooling duration for recovery. Eur J Appl Physiol. https://doi.org/10.1007/s00421-026-06141-9
15. Kużdżał A et al (2025) Dry needling in sports and sport recovery: a systematic review with an evidence gap map. Sports Med 55:811–844
16. Doege J et al (2021) Defining return to sport: a systematic review. Orthop J Sports Med 9
17. Hickey JT, Timmins RG, Maniar N, Williams MD, Opar DA (2017) Criteria for progressing rehabilitation and determining return-to-play clearance following hamstring strain injury: a systematic review. Sports Med 47:1375–1387
18. Doeven SH, Brink MS, Kosse SJ, Lemmink KAPM (2018) Postmatch recovery of physical performance and biochemical markers in team ball sports: a systematic review. BMJ Open Sport Exerc Med 4:e000264
19. Davis HL, Alabed S, Chico TJA (2020) Effect of sports massage on performance and recovery: a systematic review and meta-analysis. BMJ Open Sport Exerc Med 6:e000614
20. Pérez-Castillo ÍM, Rueda R, Bouzamondo H, López-Chicharro J, Mihic N (2023) Biomarkers of post-match recovery in semi-professional and professional football (soccer). Front Physiol 14
21. Claudino JG et al (2017) The countermovement jump to monitor neuromuscular status: a meta-analysis. J Sci Med Sport 20:397–402
22. Gathercole R, Sporer B, Stellingwerff T, Sleivert G (2015) Alternative countermovement-jump analysis to quantify acute neuromuscular fatigue. Int J Sports Physiol Perform 10:84–92
23. Plews DJ, Laursen PB, Stanley J, Kilding AE, Buchheit M (2013) Training adaptation and heart rate variability in elite endurance athletes: opening the door to effective monitoring. Sports Med 43:773–781
24. Hader K et al (2019) monitoring the athlete match response: can external load variables predict post-match acute and residual fatigue in soccer? A systematic review with meta-analysis. Sports Med Open 5:48. https://doi.org/10.1186/s40798-019-0219-7
25. Trybulski R, Muracki J, Aldhahi MI, Kurtoğlu A, Halski T (2025) Effect of cold compression and ice therapy on muscle recovery after plyometric exercise: a randomized crossover trial. J Bodyw Mov Ther 45:1010–1019
26. Trybulski R et al (2025) Effect of pneumatic and cold compression on muscle performance and recovery in combat sports athletes. Sci Rep 15:44993
27. Trybulski R et al (2025) Effects of lymphatic drainage and intense sports massage on muscle properties, damage, and function after eccentric plyometric training: a randomized controlled parallel trial. Eur J Appl Physiol. https://doi.org/10.1007/s00421-025-06101-9
28. Goulart KNO et al (2022) Fatigue and recovery time course after female soccer matches: a systematic review and meta-analysis. Sports Med Open 8:72
29. Atkinson G, Batterham AM (2015) True and false interindividual differences in the physiological response to an intervention. Exp Physiol 100:577–588
30. Booth FW, Laye MJ (2010) The future: genes, physical activity and health. Acta Physiol 199:549–556
31. Ross R et al (2019) Precision exercise medicine: understanding exercise response variability. Br J Sports Med 53:1141–1153

32. Grammatikopoulou MG et al (2022) The niche of n-of-1 trials in precision medicine for weight loss and obesity treatment: Back to the future. Curr Nutr Rep 11:133–145
33. Dingenen B, Gokeler A (2017) Optimization of the return-to-sport paradigm after anterior cruciate ligament reconstruction: a critical step back to move forward. Sports Med 47:1487–1500
34. Lettner J et al (2024) Evaluating the reliability of MyotonPro in assessing muscle properties: a systematic review of diagnostic test accuracy. Medicina (B Aires) 60:851
35. Lohr C, Braumann K-M, Reer R, Schroeder J, Schmidt T (2018) Reliability of tensiomyography and myotonometry in detecting mechanical and contractile characteristics of the lumbar erector spinae in healthy volunteers. Eur J Appl Physiol 118:1349–1359
36. Trybulski R et al (2024) Reliability of MyotonPro in measuring the biomechanical properties of the quadriceps femoris muscle in people with different levels and types of motor preparation. Front Sports Act Living 6
37. Stiver ML, Mirjalili SA, Agur AMR (2023) Measuring shear wave velocity in adult skeletal muscle with ultrasound 2-D shear wave elastography: a scoping review. Ultrasound Med Biol 49:1353–1362
38. Sendra-Pérez C et al (2023) Reliability of threshold determination using portable muscle oxygenation monitors during exercise testing: a systematic review and meta-analysis. Sci Rep 13:12649
39. Ghiadoni L, Versari D, Giannarelli C, Faita F, Taddei S (2008) Non-invasive diagnostic tools for investigating endothelial dysfunction. Curr Pharm Des 14:3715–3722
40. O'Riordan SF, Bishop DJ, Halson SL, Broatch JR (2023) Do sports compression garments Alter measures of peripheral blood flow? A systematic review with meta-analysis. Sports Med 53:481–501
41. Maia F, Nakamura FY, Sarmento H, Marcelino R, Ribeiro J (2024) Effects of lower-limb intermittent pneumatic compression on sports recovery: a systematic review and meta-analysis. Biol Sport 41:263–275
42. Versey NG, Halson SL, Dawson BT (2013) Water immersion recovery for athletes: effect on exercise performance and practical recommendations. Sports Med 43:1101–1130
43. Trybulski R et al (2024) The effects of combined contrast heat cold pressure therapy on post-exercise muscle recovery in MMA fighters: a randomized controlled trial. J Hum Kinet 94:127–146

MIX
Papier aus verantwortungsvollen Quellen
Paper from responsible sources
FSC® C105338

If you have any concerns about our products, you can contact us on
ProductSafety@springernature.com

In case Publisher is established outside the EU, the EU authorized representative is:
Springer Nature Customer Service Center GmbH
Europaplatz 3, 69115 Heidelberg, Germany

Printed by Libri Plureos GmbH
in Hamburg, Germany